THE TRAGEDY OF BELIEF

THE TRAGEDY OF BELIEF

*Division, Politics, and
Religion in Ireland*

JOHN FULTON

CLARENDON PRESS · OXFORD
1991

Oxford University Press, Walton Street, Oxford OX2 6DP
Oxford New York Toronto
Delhi Bombay Calcutta Madras Karachi
Petaling Jaya Singapore Hong Kong Tokyo
Nairobi Dar es Salaam Cape Town
Melbourne Auckland
and associated companies in
Berlin Ibadan

Oxford is a trade mark of Oxford University Press

Published in the United States
by Oxford University Press, New York

© John Fulton 1991

All rights reserved. No part of this publication may be reproduced,
stored in a retrieval system, or transmitted, in any form or by any means,
electronic, mechanical, photocopying, recording, or otherwise, without
the prior permission of Oxford University Press

British Library Cataloguing in Publication Data
Fulton, John
The tragedy of belief : division, politics and religion in
Ireland
1. Ireland. Catholic church. Relations with state
I. Title
322.109415
ISBN 0-19-827316-9

Library of Congress Cataloging in Publication Data
Fulton, John, Dr.
The tragedy of belief : division, politics, and religion in
Ireland / John Fulton
p. cm.
Includes bibliographical references and index.
1. Ireland—Politics and government. 2. Northern Ireland—
Politics and government. 3. Religion and politics—Northern Ireland. 4. Church
and state—Northern Ireland. 5. Northern
Ireland—Church history. 6. Religion and politics—Ireland.
7. Church and state—Ireland. 8. Ireland—Church history. 9. Irish
question. I. Title.
DA938.F85 1991 941.60824—dc20 90-44898
ISBN 0-19-827316-9

Typeset by Pentacor PLC, High Wycombe, Bucks
Printed and bound in
Great Britain by Biddles Ltd.
Guildford & King's Lynn

To

Frank, Kathleen, Frank
& Margaret

for all their sacrifices

Acknowledgements

I am grateful to Transaction Publishers for permission to publish in the first main section of Chapter 5 an updated and modified version of a paper I wrote which first appeared as 'State, Religion and Law in Ireland', in T. Robbins and R. Robertson (eds.), *Church–State Relations: Tensions and Transitions* (New Brunswick and London, 1986).

I would like to thank former colleagues and students of the Irish School of Ecumenics, Dublin, over the years 1973–9, especially Alan and Marjorie Falconer, Michael Hurley, and Ruth Moran. Thanks are due also to many among the membership of the Sociological Association of Ireland for their help and conversation. Particular thanks are due to two anonymous readers of drafts of the book. Thanks also to present and past social and community workers of Belfast, whose hospitality to all visitors is renowned.

My colleagues at St Mary's College, Strawberry Hill, have been especially supportive of the work, particularly Mary Eaton, Susie Jacobs, Chris Knee, Ray Lee, Beryl Mason, and the late John Kennett who is missed by us all. Thanks also to Chris Durston for his comments on the final draft of the book and Patrick MacNamara (University of New Mexico) for help with some reference details.

David Martin gave encouragement and inspired me with the necessary supply of self-confidence. To Bryan Wilson, Eileen Barker, and the members of the LSE graduate sociology of religion group, ICSR, SSSR, ASR, and BSA, many thanks also.

Most of all, a thank-you to Anne and the children, Rebecca, Rachel, and Christopher, as well as to all those dedicated and concerned people in Ireland who have taught me a great deal about the complexities of human existence. From them the lessons of living: to them peace, justice, and reconciliation, whatever the political consequence.

J.F.

Contents

Figure and Tables	x
1. *The Power Bases of Division in Ireland* The Role of Religion in Context:	1
2. Anglo-Irish Ascendancy, Protestant Subordination, and Catholic Oppression: *Ireland, 1600–1800*	25
3. Religion and the Growth of Opposing Power Groups: *Ireland on the Road to Division, 1800–1922*	53
4. States, Religions, and Ruling Ideas: *Catholic Nationalists and Protestant Loyalists Today*	89
5. Religion and Law in the Southern State: *Shoring up Monopoly Catholicism*	133
6. Schooling as Political Religion: *Systems and Counter-Systems North and South*	171
7. The Religious Boundaries of Group Reproduction: *Catholic–Protestant Marriages North and South*	198
8. Conclusion	227
References	231
Index	253

Figure and Tables

Fig. 1.1	Mass, Holy Communion, and Confession Attendance for Roman Catholics in the Republic of Ireland, 1984 and 1974	10
Table 1.1	Church Attendance of Protestants by Socio-Economic Status	12
Table 1.2	Denominational Groupings in Northern Ireland, 1961–1981	14
Table 3.1	Social Structure of Rural Ireland, c.1841–1845	69
Table 3.2	The Catholic Clergy, 1731–1871	71
Table 7.1	Mixed Marriages in Northern Ireland as a Percentage of Total Number of Marriages	199

1
The Role of Religion in Context

The Power Bases of Division in Ireland

SUMMARY

Commentators on the Ulster conflict tend to locate its causes either within the boundaries of the Northern Ireland statelet or across the waters in Britain.[1] 'The Troubles' are seen to result from the unsatisfied, Northern nationalist lust for a united Ireland, the intransigence of the Northern loyalist majority, or the lingering imperialism of Britain. The role of the Republic of Ireland is barely considered. The Southern Irish state appears simply as an aggravant to the situation, a weakling on terrorist control, but rarely an integral part of the problem of 'the Troubles'.

A number of writers consider the role of religion to be equally unimportant. The use of the terms catholics and protestants to describe the opposing groups is seen to be misleading and preference is given to the terms nationalists and loyalists. In addition, the idea that protestant fears of a catholic Ireland have any substance is easily dismissed as fruit of protestant misunderstanding and prejudice.

This book challenges such assumptions. Clearly, the British involvement is real and is part of the problems of Ireland. But, at the risk of losing readers early on, the author has omitted any

---

[1] The number of published assessments of 'the Troubles' is substantial and several articles and studies reviewing the literature have appeared, particularly: Darby 1976; Hickey 1984; Hunter 1982; Lijphart 1975; and Whyte 1978 and 1986*a*. These writers classify the explanations of 'the Troubles' either according to their economic, political, cultural, religious, or social–psychological emphasis; or according to their derivation from the beliefs of the writers—Marxist, nationalist, loyalist, or republican/anti-imperialist. Whyte (1986*a*) opts for the latter classification, but adds a group containing those contemporary academics and Marxist revisionists who tend to converge on what he calls endogenous explanations—ones which seek explanation from within the boundaries of Northern Ireland itself rather than elsewhere, such as in Britain's role in Ireland, or in events and movements in the Irish Republic.

protracted consideration of Britain's present role. The grounds for exclusion can be simply stated. The political incorporation of six-county Ulster into the United Kingdom is not the fundamental contradiction on which divisions in Ireland are based. The presence of the British state, its politics, and its armed forces, remains the second most important cause, but no longer the fundamental one. It may have been so for the sixteenth and early seventeenth centuries, but its role in this respect began to change from the time that a substantial number of protestants have lived in Ireland.

It is still true that the initiative for change in Ireland at present lies with the British state in its ability to change its relationship with protestant loyalists. But the root cause of communal violence in Ireland is the opposition of two groupings each with their own material and spiritual interests and now living in overlapping geographical boundaries of power, the protestant–loyalist and the catholic–nationalist alliances. To be a nationalist in Ireland practically means to be a catholic or former catholic, and to be a loyalist means to be an Ulster protestant. These identities are rooted in two opposing cultures for which religious belonging has great significance and into which religious belief and attitude are incorporated. Churches too are important, though religion plays a role far beyond its church identity. The basic argument of the book is that, in the shaping of the two dominant interest groups and in their endurance, religion has played and continues to play a significant role. As a constituent ideology of group dominance, religion underlies the structure of the two state powers in Ireland. It is not the principal, direct contributor to violence in contemporary Ireland, though it was also never quite such in the past. But because it forms a principal, now strong and now weak, constituent of the two antagonistic ideologies, religion enters into the existing antagonisms from which open violence stems. Consequently, religion in the concrete church forms it takes on in Ireland bears significant direct responsibility for social division and indirect responsibility for violence.

The book thus challenges a number of assumptions. One of these is the belief that violence is totally a Northern reality and that the Republic of Ireland is a peaceful, non-violent, and politically integral society, low on both civil and political crime. But while it is true that the murder, maiming, and marching mainly take place in Northern Ireland, with bombings by loyalists in the South a rare event, it is also true that Northern Ireland is the only contested territory for all the protagonists. The structures of violence and conflict are not geographically confined to the North but are located in the social world, between the two groups in Ireland as a whole, and rooted in the irreconcilability of their existing societies and aspirations.

Another assumption which is challenged is the one which is based on a popular interpretation of history. It is often said that whereas the past troubles of Ireland were related to the Reformation and colonial rule, with all the open religious oppression this entailed, the present difficulties are related to conflicts of a different kind. From within this perspective, violence in Ireland can appear to be caused by the extremists on the nationalist or loyalist sides who have been able to cash in on the vulnerability of modern society to terrorism. Again, it can be seen instead to be the result of straightforward economic injustice and political inequality, such as the clever voting arrangements which kept local government under protestant control, the continuing discrimination in jobs, and the inferiorization of the Northern catholic community as a whole.

Unfortunately, while economic and political issues are clearly central, the basis of the argument against a religious element in the conflict is not as strong on history as may at first appear. For example, it is difficult to pin-point the role of religion in Ireland in Cromwell's time, without coming up with ambiguous answers. Was Cromwell interested in pacifying Ireland or in converting it, eliminating catholicism or establishing a unified population loyal to the commonwealth? Fortunately, historians of Ireland are usually quite careful not to reduce situations to religious, political, economic, and cultural components. They are aware of the difficulty or impossibility of separating out in such a fractured way the reality they handle. Perhaps the most obvious contribution they have made to the analysis of religion in Ireland has been to clarify the development and formulation of an informal yet deep-seated accord between Irish national politicians and Roman catholic high clergy in the late nineteenth and early twentieth centuries on the respective roles each were to have in the emerging, self-governing state of Ireland. As we shall see in greater detail later, nations can have built into their continuing social-being such political–religious accords which go on to underpin their popular cultural life. However, some of the very historians who have elucidated this process occasionally let go of their insights. For example Larkin (1976) holds the view that the Irish Free State, which became the Republic of Ireland in 1937, has since its inception in the 1920s been a largely peaceful, democratic, and integral state. It will be necessary to explain at least in part what kind of peace and integration this has been.

Another assumption is the appropriateness of the methodology by which the two parts of Ireland are treated as completely separate units. State boundaries are accepted in a natural way as a baseline for understanding conflict, which then tends to be reduced to terrorism or violence caused by social unrest and trouble-makers. The fact that such state boundaries are themselves products of human political endeavours and only become given when consented to or coercively enforced is neglected. While combating the nationalist view on Ireland as a national unit, the book will accept that the island of Ireland has to be a frame for understanding the very divisions of which Irish nationalism is a part.

The framework of interpretation used in the book is derived from the work of Antonio Gramsci. This Italian Marxist and founder of Eurocommunism has three key ideas of particular relevance which can be extrapolated from his own political commitment. One is the definition of religion as a form of politics. The second is a concept of human groups as political agents, irreducible to the nation or the state. It is the notion of historical bloc. The third is his notion of hegemony, or the process of social control by the spread of particular ideas and beliefs. His analysis of religion has until recently remained largely unknown to the English-speaking world, while his work on the state is very much of current debate in Europe and Britain. With regard to the Northern Ireland state, Gramsci has been used to a degree in the analysis of political developments within the dominant protestant–loyalist group by Bew, Gibbon, and Patterson (1979), Bew and Patterson (1985), and O'Dowd, Rolston, and Tomlinson (1980). Both sets of authors help us to understand the complexity of political processes and control within the dominant population of the Northern statelet since 1922. Both are also deeply aware of the stark and unavoidable reality of protestant–loyalist power and identity, however internally divided its people may be. But none of the authors expands on what sectarianism is as a form of culture. Because the notions of religion and bloc as used by Gramsci are still little known, they are

The Role of Religion

clarified in the final section of the chapter. In addition to Gramsci, bits and pieces from other sources are used to extend the theoretical framework, such as Buckley 1984, which examines Ulster protestant interpretations of history and everyday life.

Given the range of material covered by the book and yet the centrality of a single problematic—the role of religion in divisions in Ireland—it has been necessary to use a variety of investigative techniques. This chapter continues with a critical section on the extent of 'good' religion and 'bad' sectarianism in Ireland. The final part of the chapter outlines the role religion is seen to play once one uses the conceptual framework suggested above.

Chapters 2 and 3 trace the relationship between religion and division in Ireland from the seventeenth century to the early twentieth century and examine the formation of two opposing groupings in Ireland. Chapter 2 examines the role of religion in the formation of the protestant alliance, in its period as part alliance, part caste relationship between the dominant ascendancy and subordinate dissenters. Chapter 3 concentrates on the role of religion in the formation of the catholic–nationalist alliance, particularly from the failure of the united Irishmen in 1798 to the latter part of the nineteenth century, as well as on the emergence of the protestant loyalists from the previous Anglican-protestant grouping. To produce a brief account of these historical developments in only two chapters, it was necessary to read in a wide-ranging way on Irish history, relying on those works which attempted a more sociological analysis of interactions between religion and political–economic processes. Those familiar with the role of religion in modern Ireland may wish to omit Chapters 2 and 3 and move straight on to Chapter 4, which completes the historical analysis with an account of the role of religion within the protestant–loyalist and catholic–nationalist cultures of dominance in the twentieth century.

The final three chapters present three case-studies on the present relationship between religion and group solidarity especially among catholic nationalists. The studies show the involvement of monopoly catholicism in the construction and maintenance of oppositional group power. Chapter 5 outlines the religious input into the development of constitutional law in the Irish Republic and its translation into political practice. A comparative analysis of the Irish constitution of 1937 and papal documents of the period on social ethics is first presented. In researching these, use was also made of Whyte's substantial account (1980) of the development of church–state relations in the Southern Irish state from 1922 to 1980, particularly the sections covering 1922 to 1952. The chapter is then completed by an analysis of constitutional developments from 1980 which have tended to reinforce religious power in the Republic rather than modify it substantially.

Chapter 6 examines the political–religious process of power and control in the sub-sphere of the education system. The extent of Roman catholic church power and its relationship with the state within the framework of the catholic–nationalist alliance is seen to be particularly evident in events concerning the integrated schooling movement. Data for this section were obtained in part from documentary analysis, particularly the newsletters of the All Children Together movement based in Belfast, and the Dalkey School Project based in Dublin. Much was also gleaned from the writer's role as an observer of these movements in their formative years as well as from interviews with some of their leading members.

Chapter 7 examines the role of catholic clergy and doctrine in controlling the endogamy of catholic nationalists and the effects of such control on the antagonism between the two groups up to the present. This is the issue of the social significance of mixed, protestant–catholic marriages throughout Ireland. The topic is important because it shows the impact religion can have on the most basic of social structures. The subject was researched by consulting official documents of catholic and protestant churches, discussing the issue with theological specialists, following the debates in the appropriate journals and newspapers, and through contact with counsellors and groups working in the area.

The idea of writing a lengthy conclusion on policy was considered and then abandoned as too ambitious. Brief reflections are made instead. One thing is clear: because culture is something which can be shaped by preachers and teachers as well as by economy and events, the future of the island of Ireland and its people or peoples lies also in the hands of its religious, educational, and political establishments. Whatever the state boundaries existing in Ireland, the analysis in the book points to the need for a change in political theology by the churches. It also implies that educators and politicians parallel them with a similar reformation.

GOOD RELIGION AND BAD SECTARIANISM: SAVING RELIGION FROM DISREPUTE

Until recently there has been little in the way of deliberate or detailed assessment of the relationship of religion to 'the Troubles', or to conflict in general in Ireland. In fact some approaches seek to correct the popular understanding of catholics and protestants being precisely that and get rid of religion as a contributory factor altogether. The clearest statement of this position is McAllister 1982. In an Irish mobility survey of 1973 which was never published, John Jackson collected some information on religious belief as well as on political attitudes. McAllister uses this information to construct a measure of religious commitment and tries to see if having more or less of this religious commitment is in any way related to people's political attitudes. He assumes that political attitudes which blame the other side can be considered causes of 'the Troubles'. He also assumes that if people show strong religious commitment at the same time as showing negative political attitudes, then some relationship between religion and 'the Troubles' could be said to exist. In fact he finds little relationship between the political attitudes of catholics and protestants and their religious commitment. He thus concludes that there is no relationship at all between religion and the Northern Ireland conflict.

McAllister's conclusions are flawed in several ways. He takes no account of the role which religious élites as opposed to the general population can have on the political process. Several authors, including Bruce (1986b), MacIver (1989), and Smyth (1987), have shown the decisive political role which protestant fundamentalists, particularly the Paisleyites, have played in oppositional politics in Northern Ireland. Also, in the absence of other material, McAllister was forced to use very limited measures of religious commitment. For 'belief in the supernatural', he used questions on belief in the Devil, hell and the afterlife, belief in divine providence, and belief in the historical truth of Bible miracles. These beliefs are hardly likely to be the ones with a politically sharp cutting edge in the modern world. Finding no relationship between these beliefs and politics is not the end of the matter. A relationship remains possible between other deeply held religious beliefs and the politics of participants in the conflict. For example, Hickey (1984) recognizes as a fact of history the drive of catholicism to control culture and public morality in society. He sees how intolerant such a religious attitude is towards the beliefs and freedoms of others. Hickey has a very different assessment from McAllister, because he has a different way of perceiving the religious–political link.[2]

If one is looking for the public attitudes of writer-politicians rather than social scientists on 'the Troubles', religion only appears to figure as an important ingredient for the politicians of the Democratic Unionist Party, such as Ian Paisley (1966), and for two well-known Southern academics and past politicians, Conor Cruise O'Brien (1974) and Garret FitzGerald (1972). Other political figures, at least in their public statements, consider the conflict to be a matter of politics more narrowly conceived, as might be expected.

If one looks at the views of religious leaders, the issue becomes more complex. Protestants appear more able to accept that religion continues to play a role in the conflict. This ability may well be based on a greater protestant readiness to view the church on earth as capable of evil, intentional or otherwise, whereas the dominant catholic theology finds it hard to accept sin within the church. Its clerical leadership is very sensitive to criticism which might mar the church's public image. There is therefore little theory of bad religion except of that religion which other churches and religions might possess. In fact, for some time, Irish Roman catholic bishops, eager to defend their own church and its social policies, such as the maintenance of catholic schools, spoke out to deny that religion played any role at all. This position has attenuated of late. After a unit of the provisional Irish republican army bombed Enniskillen's main square on the occasion of the annual Remembrance Day service in November 1987, with the loss of sixteen lives, Cardinal O Fiaich, Roman catholic primate of all Ireland, publicly begged for forgiveness for any failure on the part of catholics and their church to prevent the violence. But whether this admission amounts to a genuine theological conversion on the nature of the church and is to inaugurate a new area of political ecumenism remains to be seen. In March 1989, O Fiaich defended an Irish nationalist's right to honour the heroes of the Easter rising of 1916 as part of the belief and heritage of the national tradition. This is hardly an attitude indicative of revisionism (*Irish Times*, 18 Mar. 1989).

The view that religion is the enemy of 'the Troubles' is the first

---

[2] A fuller critique of McAllister can be found in Fulton 1988*b*: 6–9.

The Role of Religion

matter which must now be dealt with. My suggestion is that this view finds support in a particular understanding of what religion actually is as a social reality. It is this understanding which hides religion's contribution to Ireland's divisions from many catholic and some protestant minds.

Good Religiosity?

The widest amount of empirical data on religion in Ireland comes from a source which has frequently assumed the positivity and essentially good character of religion, that is the institutional church, and particularly the Roman catholic in its *sociologie religieuse*. Here, religion has been treated as a measurable quantity of an intrinsically good substance, of which there can be more or less. In Ireland, when negative elements related to religion appear, both religious and secular investigators tend to use the term sectarian to describe the acts and attitudes. The term religious extremism is used only in relation to militant protestant fundamentalism. Sectarianism is a term also used to describe what are frequently considered non-religious phenomena, such as ethnic and class divisions. We shall take a look at the presence and absence of this good religion and bad sectarianism in Ireland. Such a strategy will help us take stock of the strength of religious practice. It will also permit preliminary examination of what appear to be the contradictory actions of its religious people and of the seeming inability of the conceptualization religion/sectarianism to make sense of these actions.

Advanced capitalist development or modernization, a process viewed as intrinsically secularizing by some (Berger, Berger, and Kellner 1974), coexists throughout Ireland alongside the highest levels of traditional religious practice among both catholics and protestants throughout the whole of Europe. In the twenty-six counties of the Republic, 94.5 per cent of its three million inhabitants were listed in the census of 1971 as catholic. Of these, by 1974, 95 per cent claimed to believe in God, 93 per cent expressed a personal need to believe in God, 99 per cent perceived church membership to be important, and 65 per cent experienced some form of God's presence in their lives. Ninety-one per cent went to Sunday mass and 23 per cent went to mass more frequently. Ninety-two per cent made annual communion and 30 per cent even weekly. Forty-six per cent attended confession monthly and 80 per cent practised daily prayer.[3] Figure 1.1 shows the minor changes which appear ten years later in a follow-up study (Breslin and Weafer 1985: 10). Fogarty, Ryan, and Lee (1984) provide similar results.

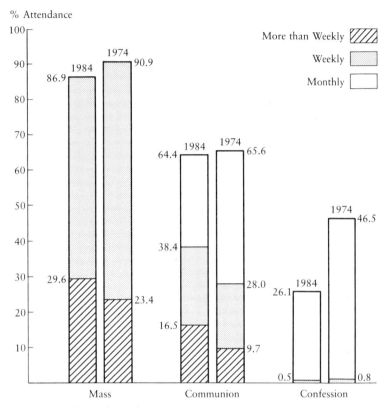

Source: Breslin and Weafer 1985: p. vii. The data in the surveys cover a national sample (26 counties) of the 18 + age group. Findings from the 1974 survey can be found in Nic Ghiolla Phadraig 1980.

FIG 1.1. Mass, Holy Communion, and Confession Attendance for Roman Catholics in the Republic of Ireland, 1984 and 1974

However, the changes which have occurred can be interpreted either as a strengthening or as a weakening of formal religiosity.

[3] The data are valid for the over-18 age group. See Nic Ghiolla Phadraig 1980; also MácGréil and O'Gliasain 1974 with comparable data for Dublin.

The Role of Religion

Supporters of the secularization–modernization thesis will note that

the highest levels of religiosity were recorded for such traditional groups as those brought up or presently living in rural areas, the farming sector, those whose education finished at primary school, those who never lived abroad, the older respondents, women, those who did not consume foreign media. (Nic Ghiolla Phadraig 1981: 364)

City dwellers, factory workers, and those involved in business and commerce showed lower, if still high levels, and the unemployed lower ones still (Fogarty, Ryan, and Lee 1984). The picture in Northern Ireland for catholics is only mildly weaker. For example, in the Limavady area, church attendance is only 5 per cent less (Hickey 1984: 129).

The small Church of Ireland community in the Republic, 4 per cent of the Southern population, also has strong church attendance and strong parochial and synodal participation rates (K. Bowen 1983). In Northern Ireland, the church attendance of protestants, including Church of Ireland, for the city of Belfast has declined from 46 per cent to 39 per cent since the civil strife began in 1968. But in a typical Ulster town, some 65 per cent of protestants attend Sunday worship at least once a month. Within Northern Ireland as a whole, the figure is likely to be similar, as can be seen in Table 1.1. Supporters and members of protestant paramilitary groups are frequently unchurched, though this is less the case on the catholic side.

The situation among young people is less clear. Lower rates of belief and religious concern are generally recorded for the North (Turner, Turner, and Reid 1980: 9). For the South and its catholic youth, some as yet unpublished research by the writer indicates wavering belief about the existence of God for about one in four 16- to 18-year-olds, while this remains a simple fact for their parents. If one notes that a movement in this direction, though not as marked, among the 18–25 age group has been pointed out by Breslin and Weafer (1985), one can suggest that there is a shift from traditional convictions, and that this is occurring in the early teens. Even so, it is doubtful if complete dissociation between religion and Irish identity would easily occur. There are few alternatives in Ireland: a gradual weakening of ties and of intellectual content, or the non-religious, part ethnic, part Marxist, and strongly

TABLE 1.1. Church Attendance of Protestants by Socio-Economic Status

Attendance	Status					
	A	B	C1	C2	D	E
Once or more a week	44.4	50.0	47.5	28.0	34.1	35.5
Less than once a week but more than once a month	14.8	23.4	20.0	22.2	20.4	12.9
Less than once a month or never	40.8	26.6	32.6	49.8	45.5	51.7

Note: 'Class A includes senior professional[s] and executive[s], B includes lower professionals such as school teachers and university lecturers, and also middle managers, C1 encompasses routine white-collar workers (for example bank clerks and telephonists), while C2 mainly contains skilled manual workers. D signifies unskilled manual workers, and E is a residual category for pensioners, invalids and casual workers. For further details see Moxon-Browne 1983:199' (Wallis, Bruce and Taylor 1986: 6).

Source: Northern Ireland Attitude Survey, conducted by E. Moxon-Browne in 1978, most of which is published in Moxon-Browne 1983. The table appears by permission of Moxon-Browne in Wallis, Bruce, and Taylor (1986: 6).

nationalist allegiance of some within the provisional movement. The small numbers of atheistic humanists are likely to remain an intellectual minority.

Add to these factors the high numbers who still join religious orders, both male and female, on the catholic side. The rapid decline of the late 1960s restored 1950s' levels, though the decline is continuing. Add also the strong missionary activity abroad of all the denominations, coupled with a certain amount of renewal, parish and charismatic. One can even include the religious political involvement of Irish catholic missionaries in such places as Nicaragua, San Salvador, and the Philippines, a strong justice and peace movement reaching up to the hierarchy within Ireland itself, and a lively, if intellectually led, ecumenical movement, with its own research centre and student groups north and south of the border.[4] The resultant picture in many respects is of a modern, lively Christianity, a quite strong, good religiosity.

---

[4] The Irish School of Ecumenics, Milltown Park, Dublin.

Bad Sectarianism and Violence

The difficulty with this view of Irish religion, using as it does the traditional indices of *sociologie religieuse*, is that it is far removed from other sorts of things with which Irish men and women get involved or sympathize. Some intermediate-type data can help. For example, there is a considerable degree of authoritarian attitude on the catholic side and this is associated with high levels of religiosity (Nic Ghiolla Phadraig 1981). This suggests rigid attitudes which, in the minds of one school of social psychologists, could lead to violent actions (Adorno 1950). Equally, as indicated in Table 1.2, on the protestant side fundamentalism is a strong force in Ulster and might well be on the increase. In particular, the Revd Ian Paisley's Free Presbyterian Church, with its particular emphasis on the salvation experience through Jesus Christ alone—the 'alone' indicating above all the rejection of the evil teachings of the Roman church—has grown from 1,300 in the early 1960s to over 10,000 in more recent times. Paisley takes to the streets when there is any indication that the British government is weakening in its intention to support protestant hegemony in Ulster. He and the remainder of the fundamentalist leadership of the Democratic Unionist Party have supported those who have felt it necessary to prepare in a military way, lest Britain betrays Ulster protestants (Bruce 1986*b*; MacIver 1989). The grim, yet cheery toughness and boldness of the marching tradition is likely to have a link with Old Testament religious conquest and warriorship, though Paisley himself has appeared somewhat ruffled when people have actually got hurt.

Fundamentalist politics can be seen as a minor religious aberration, with mainstream religion remaining basically good or at least irrelevant. But then we come to data which are barely reconcilable with this view. On the catholic side, a high level of disobedience to the hierarchy appears to occur on the issue of support for the men of violence. In Ulster, it can no longer be asserted that only nominal catholics get involved in the provisional movement or that violence has no credibility within the catholic–nationalist community as a whole. At first, it may have been possible to interpret the electoral support for the hunger strikers of 1980 as support for civil rights. But it is difficult to interpret the electoral success of Sinn Fein in the Assembly election of 1982, the local election of 1985, and the Westminster elections of 1983 in

TABLE 1.2. Denominational Groupings in Northern Ireland, 1961–1981

Denominations	1961	1971	1981
Roman Catholic	497,547	477,919	414,532
Presbyterian	413,113	405,719	339,818
Church of Ireland (and all Anglicans)	344,800	334,318	283,975
Methodist	71,865	71,235	58,731
Brethren	16,847	16,480	12,158
Baptist	13,765	16,563	16,375
Congregational	9,838	10,072	8,265
Non-Subscribing Presbyterians or Unitarians	5,613	3,975	3,373
Reformed Presbyterian	4,163	3,904	3,202
Salvation Army	2,028	2,063	1,790
Elim Church	1,768	2,269	3,413
Jew	1,191	0,959	0,517
Free Presbyterian	1,093	7,337	9,621
Society of Friends	1,067	1.019	0,764
Church of God	1,001	1,018	1,294
Protestant	0,980	4,423	14,318
Pentecostal	0,959	1,418	2,156
Christian	0,253	3,691	8,695
Agnostic (fewer than)	10	1,200	1,171
Jehovah's Witness	0,729	1,041	1,214
Latter Day Saints	0,371	0,975	1,067
Free Methodists	0,080	0,619	1,103
Undenominational or Unsectarian	0,789	0,852	1,217
Not Stated	28,418	142,511	274,584
Atheists	0,087	0,512	0,730
Total population (including groups not listed above)	1,425,042	1,519,640	1,418,959

Note: The recent round of 'the Troubles' began in 1969. Fear of revealing one's religious identity is *one* explanation for the growth in the 'not stated' category. Of course, there may be others.

Source: *Census of Northern Ireland* for 1961, 1971, and 1981 (small counts omitted)

the same vein. In all cases, Sinn Fein candidates were elected or voted for in substantial numbers, about 40 per cent of the catholic–nationalist vote. The slogan for these elections was 'the rifle in one hand and the ballot box in the other'. Explanations in terms of temporary frustration with the political process appear too pat and take away from the rationality of the actions. The first endorsement of violence was at the end of October 1982, the second on 9 June 1983, and the third in May 1985. In the election of January 1986, there was a considerable decline in the Sinn Fein vote. It appears to have been directly related to the Anglo-Irish agreement of November 1985 and may indicate a wish to pursue constitutional politics when these are seen to be successful. The Sinn Fein vote has since remained relatively high at 25 per cent of the Northern Catholic population and may go higher again if the peaceful road to power again fails to make some progress. In the South also there is and has been considerable support for the path of violence. This is not just a matter of past history but is present in the widespread belief that violence does succeed in the case of the North. For example it is the view of 35 per cent of Dubliners in Mác Gréil's survey (1977) that violence has been necessary for the achievement of non-unionist rights. To the minds of this significant minority it would appear that violence has succeeded in making Britain move against the protestant–loyalist domination of Northern Ireland and that such violence was morally justified.

With regard to the protestants of the North, a similar pattern can be detected. The belligerent attitude of a great part of the protestant urban working class, the strong words and sometimes actions of the DUP leadership, the parading of symbols of antagonism, open hostility of language, and general provocation of the catholic–nationalist minority, all under the banner of 'no popery' and 'a protestant state for a protestant people', add up to a powder keg for which a single reprehensible action by nationalists can become a lighted fuse. Several leading protestant clergymen are in politics, and all of them support a policy of no compromise with the Irish Republic or with the nationalists within the Northern statelet. They are against any attempt to share power with catholics or nationalists unless they take an oath of allegiance to the British crown. Clergy of the free presbyterian, presbyterian, and Church of Ireland communities, are often members of the Orange order, the society which unites a unique blend of patriotism to the British crown with a religious dedication to protestant values. A recent grand master of the order, Martin Smyth, is a presbyterian minister. It is generally recognized that the inner core of the order, which is modelled on the order of freemasons, is made up of more religious persons, promoted for their upstanding morality and dedication to the protestant faith. Of all the organizations opposed to co-operation with republicans and catholics, the Orange order has certainly been the most powerful and popular in the long term. Though the men of violence among the protestants are frequently non-practising in religious terms, fellow religious protestants feel sympathy toward them. They are seen to be defending the cause in a necessary way. Then there is the participation of fundamentalists in street politics. The reluctance of many of these to endorse specific acts of violence or raise a hand in physical attack does not remove the proximity to violence which the pursuit of such street politics entails. Again, there is a widespread belief in the effectiveness of violence.

At the end of the day, for the majority of catholics and protestants in Northern Ireland, and a significant number of Southern Irish catholics, the men of violence are 'the boys' or 'the lads'. In the ghettos of Belfast and Derry, mass attenders sometimes become gun minders when such action becomes seen as necessary to save the skin of one of the lads from the British army or Ulster police. Of course, it is possible for violent persons to remain religious within their own conscience whatever the judgement of others, not because they are mentally distorted, but because their shared, practical, and religious view of the world allows them so to do. This is true for a number of politically violent people and their more passive supporters in Ireland.

The beliefs, commitments, and experience of catholic and protestant in Ireland clearly appear to be far more complex than the well-worked-out social–psychological concepts of *sociologie religieuse* would suggest. But then those measures of belief and behaviour were mainly constructed to measure the beliefs and behaviour of the church faithful against churchmen's own conceptualization of what the good Christian life was. As such they tended to emasculate rather than fill out what religion might mean in the world of conflict and politics. To deal with the latter it is necessary to understand the experience of religion at the level of its operation on the streets, the influence of the institutional church on its faithful, and the linkages of religion with state beliefs and ideologies. In other words, we need to explore the contribution of religion to the overall structures of power in Ireland, North and South. It is also important to note that the levels of support for violence cannot be reconciled with the levels of good religiosity unless they are both attributed to a number of the same people.

The Role of Religion

Some link between religion and politics is recognized by many in Ireland. In fact, protestants are joined by Garret FitzGerald and Conor Cruise O'Brien in finding the Roman catholic church in the Republic to have political power and to exercise it negatively. They find catholicism so invasive of the public sphere as to constitute both a secular and religious threat: to civil liberties over rights of divorce, contraception, and the like; and to what protestants view in general as their religious heritage, which includes the right to a certain freedom of moral action in their personal lives (Moxon-Browne 1983: 39).

To understand religious power in Ireland we need a set of concepts, a framework of interpretation. At the same time, we need to overcome the established categories of positive religion and negative sectarianism. Only in this way will it be possible to clarify the role of religion in Ireland's divisions, both in the past and in the present. To this task of framing the problem and developing the appropriate concepts we now turn.

BLOC POWER, HEGEMONY, AND RELIGION IN IRELAND

Shaping the Question

The terms unionist, protestant, and loyalist express some of the subtle differences in emphasis among the majority of Northern Ireland's population. But they are not always attributable together to specific individuals. Instead, the term protestant–loyalist identifies the majority as a unified social group or alliance. Even if all members of the group are not Bible-thumpers, 40 per cent of them follow the particular line which Paisley and the Democratic Unionist Party articulate so well. Most of the remainder follow the official Unionist Party and while they reject Paisley and his street politics, they still remain sympathetic and committed to the goal of a protestant-dominated Ulster, whether they wish to remain British or not.

The term catholic–nationalist, while identifying the dominant political and numerical group in the Republic of Ireland, also identifies three-quarters of the northern minority, the other quarter being mainly catholics who are more indifferent or even opposed to Irish nationalism. In addition, whereas Northern protestants are divided into a number of churches and sects, catholics belong to one church which appears to have special links with the Southern

Irish state. In matters of public morality other than violence, Irish catholics have been particularly obedient to the church in recent years. Also there has been a significant degree of institutionalization of catholic morality in the Irish constitution, with subsequent limitations on freedom for protestants and other believers, issues which will be covered in depth in Chapter 5.

Both groupings, catholic–nationalist and protestant–loyalist, appear to be locked into similar processes. They possess a degree of coalescence of identity on the basis of powerful beliefs and practices rooted in religion, culture, and political orientations. We could opt immediately to use the terms 'ethnic groups' to identify the two sets and leave it at that. But because the label might blind us to the processes modifying or strengthening group solidarity and their mutual opposition, it is perhaps better to avoid labelling for the time being and to prise open the ethnic box so as to see something of its real, social content.

Historical Blocs, States, and Civil Societies

It is perhaps hardly surprising that the Gramscian term historical bloc to describe the protestant–loyalist grouping in the North has come to be used by some writers (Donnan and McFarlane 1986: 25; Gibbon 1975) intuiting a framework needed but not yet systematically applied or developed. The term may also be fruitfully applied to the catholics throughout Ireland as well, despite their separation by state boundaries and the fact that they serve the interests of two different political masters. When Gramsci (1971; 1975) invented the term in the 1920s, he was addressing the problem of how to understand the totality of relationships between people in modern societies. He concentrated on those social processes which most bound people together and tried to marry his perceptions to a Marxian-type understanding of the nature of power. He found the traditional division which had currently become popular—that between economic structure and cultural superstructure, between the relationships of production and the institutions of the state, the family, religion, and culture—far too simplistic and in need of review. Instead, while accepting the primacy of work and production in shaping the structure of power relationships within and between societies, he conceived of the complexity of the relationships in the following way.

Firstly, the so-called superstructure of the cultural life was seen to interact with the productive sphere so intimately as to belie any total separation. Culture and religion were seen to be creative, and therefore another productive form.[5]

Secondly, and most importantly for the analysis of division in Ireland, he captured two sets of linkages, one between economy and culture, the other between the state and the remaining institutions, by creating the term historic or historical bloc. This could be translated simply as social bloc or bloc of interests, though the terms grouping and alliance will frequently be used in the present text. The term historical indicates that the realities of human history must take precedence over any type of theorizing. Events and the ongoing dynamics of social life are the prime stuff or materiality of all reflection. The term bloc refers to the process of binding people together to form a group or society of some kind, even if their interests are divergent or contradictory. Key interest groups are not simply those which divide into slave and master, lord and serf, capitalist and worker—the opposing economic groups we recognize in the economic history of our societies—but also other groups that one finds enmeshed in these economic orders, such as the clergy and merchants of medieval society. Thus, the notion of bloc extends the over-simplistic historical categorization of periods solely in terms of class, giving it a greater sophistication. The whole term, historical bloc, refers to the fact that one of the prime characteristics of such groupings is that they endure for extensive lengths of time. They are so important to the nature of events that they style the span of the time of their dominance into historical periods, such as the classical or the feudal.

Thirdly, and again most importantly for Ireland, Gramsci did not identify either geographical realities or states as the sole reference points for studying social relations, but stressed the centrality of bloc power. Though Gramsci actually studied historical blocs which have their own state structures and civil societies, he provides a framework whereby one can hold the one or the other part in suspension.[6]

---

[5] Gramsci's view on production at this point is similar to Marx's (1971) own early notion of the human.

[6] Here I am following Portelli (1974), whose interpretation permits a more immediate grasp of the Irish situation than Christine Buci-Glucksmann's (1980). Her analysis begins with the state, probably the French, and finds Portelli's emphasis

Gramsci devotes the greater part of his writings to superstructural aspects of power and social life rather than to the economic, except when he deals with ideologies of production such as Fordism and scientific management. He does not see the state as simply an arm of the bourgeoisie, but as generating its own power base. Also he does not see the state as the key source of power or sole complement to the power of the ruling class. Rather, there are two institutional spheres interacting both with each other and the economic: namely, civil society and political society. The former comprises the family, religion, leisure institutions, high or upper-class culture, and popular culture, which consists of the often conflicting traditions and mores of the more or less dominated populace of a society. Political society is the institutions of the state as we would normally understand them: central and local government, welfare, and the forces of law and order, or army, police, and the courts. But the power any single institution or group may wield within the body politic depends primarily on its particular relationships within the bloc.

The assertion of the semi-autonomy of civil society from the state raises the suggestion that in Ireland the two civil societies which are populated by the blocs of catholics and protestant loyalists do not correspond to the two state structures. Nationalist values in Northern Ireland continue to motivate the catholic–nationalist remnant in varying types of resistance to the state, from ritual parades to violent acts. At the same time, the recognition of civil society itself as a source of power allows one to say that the states of Ireland themselves are apparatuses of power which are based not on themselves, but on the group alliances which have given and continue to give rise to them. For example, the Roman catholic church in Ireland and the Orange order, institutions forming part of the two opposing blocs and civil societies, are organizations which have had a considerable effect on the structuring of decision-making within each of the states of Ireland. Understanding the internal power which structures these blocs or groupings will give us an improved grasp of the role of religion in the production of unity and opposition in Ireland.

At the heart of historical blocs are alliances of those groups or

on the historical bloc to be at fault. Adopting the looser framework of Portelli does seem to make greater sense in the present context, whatever the academic arguments over what Gramsci actually meant.

classes in whose interests the present mode of production and ordering of society operates. A bloc in a capitalist-type society might be dominated by capitalists, managerial élites, aristocracy, military leadership, the church, and the state bureaucracy, with the last's abiding interest in the continuation of rational procedures based on the status quo and on improvements which benefit the bureaucracy itself. Such alliances can vary from time to time, some groups entering and some leaving the alliance. The fundamental group or groups will tend to endure, and form the central and distinctive feature of the bloc as such, so long as it or they succeed in maintaining their interests. There will also be modifications in the fundamental group brought about by a variety of historical developments and the growth of technology.

Hegemony, Coercion, and Religion

The degree of success of the dominating groups within the bloc will depend on the extent to which they incorporate the groups furthest removed from power and succeed in subordinating these groups' interests to its own. The interests of different groups in society will only sometimes coincide, as in war against an invader. Otherwise they are more likely to conflict openly if not somehow controlled. This can be done in two ways: either by coercing subordinate groups, that is using force; or more successfully by presenting the ruling group's own interests in a universal way, that is in such a way that those interests come to be seen by the subaltern or subordinate classes as their very own interests. This is why Gramsci names the two types of social control in society as coercion and hegemony.

He uses the terms specifically in the context of the modern state, with geographical boundaries successfully defended from outside aggression. If one looks at the situation in the capitalist-democratic state, then for most of the time it is not under siege from groups within its boundaries and is therefore not in a state of crisis. This appears to be the normal situation for those living within the state, and is seen by Gramsci to be social control by hegemony: consent to the status quo on the basis of commonly shared values and beliefs in the rightness of the social order. It is in periods of social unrest that coercion comes to the fore.

Hegemony operates mainly through civil society—family, church, education sector, as well as political parties, trade unions, and clubs. However, it is also true that state culture provides significant elements of hegemony, such as the monarchy in Britain and pride in the nation-state. The institutions of civil society tend to carry values which bind the subordinate classes to the political domination of the ruling groups. Central institutions of civil society are usually family and church. Historically religion has played a key role in providing hegemonic beliefs for dominant alliances. In medieval Europe, religious leaders have even played the role of 'religious and organic intellectuals', that is organizers of the body politic.[7] For Gramsci, religion has a direct impact on the struggle for power in society. It is 'an active conception of the world' which prepares people for a political role in society. This role can either be one of passivity, especially in front of an alien dominant group, but only so long as that group remains dominant, or of activity through reform or revolution. All active conceptions of the world, the most powerful of which is socialism for Gramsci, contain elements of belief, commitment, culture, and power. Even when the society in question is not a theocracy, religion can play a crucial role in the structure of control or change. Religious institutions and not just religious movements can be involved in the political process, though this activity can be indirect and not always obvious.

The Irish Experience

We have here a fairly good model of social relationships to apply to Ireland. The protestant–loyalist bloc or alliance originated in a previously island-wide protestant alliance which developed in the seventeenth century and was shaped around the landowning ascendancy. At the turn of the nineteenth century, the structure of the alliance was modified in the face of the threat of a catholic–nationalist take-over of the whole of Ireland, and became dominated by economic and religious interest groups within Ulster alone, cutting out protestants in the rest of Ireland. It remains the dominant grouping in the North of Ireland today, though not without its internal contradictions, struggles, and fissures. It is a grouping split between those loyal to themselves and those loyal to Britain; and there would clearly be changes in its structures of dominance were British rule to disappear.

The other bloc in Ireland emerged from a subject population

[7] For a brief presentation of Gramsci's historical sociology of religion, see Fulton 1987a.

whose sense of identity at the time was mainly based on its catholic belief. Before the nineteenth century, that population hardly existed as a political adversary as it was wholly subordinated to ascendancy interests largely by coercion and partly by passivity. But by the end of the nineteenth century it had become a much more integrated grouping, dominated by a separate set of economic, political, and cultural interests, competing with its protestant masters for total control. In partnership with its church leadership, this catholic–nationalist alliance eventually replaced protestant domination within three-quarters of the territory of Ireland. With independence and the construction of its own state form throughout that territory, part of the grouping's constituency, the Ulster catholic nationalists, was abandoned to the mercy of the protestant loyalists of the North and was cut off from the experience of this freshly gained autonomy.

Now the catholic–nationalist alliance still seeks to a significant extent domination over the whole island, and the protestant–loyalist alliance seeks to maintain its domination only over a flexible area of North-East Ireland. Both tasks are impeded coercively as well as ideologically by the opposing grouping or by some interest groups within the opposition. Protestant loyalists have been incorporated into the British historical bloc, but they have now been loosened actively from this alliance by Britain. Britain's interests no longer require the union with Northern Ireland. The protestant–loyalist grouping thus finds itself in a post-colonial phase, in search of an identity and justification both for its dominance and for its very existence. The two opposing groups indigenous to Ireland are both based on capitalist-type economic activity and class structure, though the development of both has been of different and uneven kinds. Both bind into their dominant philosophies tenets derived from liberal economics and morals as well as from Christianity. On the catholic–nationalist side, the religious elements have until recently been provided by the taken-for-granted structures of the catholic monopoly state form. On the protestant–loyalist side, what binds protesters together against the catholics, rather than what divides them into Anglicans, dissenters, and methodists, are key religious elements woven so tightly into the fabric of free enterprise as to be almost indistinguishable. In a sense, one has the primordial philosophy of the protestant ethic from which both religious and secular views of individualistic economic activity are, at least in part, derived. Because catholic nationalists have tended to hold an ethnocentric view of history and claim the island as a whole both as their birthright as well as their democratic right, their hegemonic philosophy constitutes, at the same time, a nationalist philosophy which is theoretically blind to the existence of protestant loyalists as an opposing group of economic and cultural interests. The protestant–loyalist philosophy is less clear in its content, being the fruit of a colonizing experience which involved the subordination and not the extirpation of the enemy. Key groups within this alliance oppose the development of a national ideology on principle, as will be seen.

In other words, there is a structural conflict in Ireland between two major historical blocs. They hold to a mutually exclusive awareness. Their intolerance of each other comes in part from the identity and power aspirations of the other. The awareness is underpinned by their respective interests and is coupled to distinct religious beliefs. This coupling, though occurring in different ways for each group, is so intimate that the experience of religion and ethnic identity is frequently the same. It will be argued that, in the case of both groups, religious identity is socio-political, and openly or secretly justifies the control and subjugation of the opposing group. Traditional church forms of religion—Calvinist protestantism and monopoly catholicism—have played and continue to play a significant role in the hegemony of both blocs, and permit some shift in internal class and group alliances without disturbing the solidarity of the blocs overall.

In this book, the explanatory framework for understanding division, politics, and religion in Ireland is thus a holistic as well as historical one. That is, explanation is found not in any single element but in a complexity, with the relationship between the elements being of crucial importance. It is the way particular beliefs and cultures relate to material concerns, the state, civil society, and historical events which has to be understood and which is also the most difficult to grasp. Of course, the understanding produced by this theoretical perspective has to be plausible. Here we are caught within the paradigm: using the frame to interpret the conflict in Ireland helps shape our understanding of the various realities themselves. Hence, the self-fulfilling prophecy occurs. What the reader has to judge is if it makes more sense than previous explanations offered.

2
Anglo-Irish Ascendancy, Protestant Subordination, and Catholic Oppression

Ireland, 1600–1800

Chapters 2 and 3 trace the historical relationship between religion and division in Ireland. They concentrate on the development of the two opposing alliances and the role religion plays in the process. Understanding the relationship of popular religion to the process is largely underdeveloped, as historical interest in the area is a late arrival on the scene (S. J. Connolly 1982; S. Connolly 1985; Miller 1975; 1978a; 1978b; Hynes 1978). It will be seen that the protestant–loyalist grouping developed from a pre-existing Anglo-protestant alliance (Ch. 2) but in interaction with a newly forming catholic–nationalist alliance (Ch. 3); and that the present dominance of the two groups in two geographically ill-fitting states is a key phase in the continuing process of conflict (Ch. 4). Throughout this process, the interaction between religious and other cultural motivations will be seen as crucial. They provide elements for the construction of the hegemony of the emerging and opposing groups.

THE ORIGINS OF A PROTESTANT BLOC IN IRELAND AND ITS MATERIAL INTERESTS, 1530–1700

Henry VIII's reformation of the Irish church did not lead to the development of religious strife in Ireland. Despite the dissolution of the monasteries, the people remained attached to their traditional religion, an indication of the strength of religion's popular base. Even those aristocrats and feudal lords who remained loyal to Henry in a largely unpacified and unconquered Ireland differed with him on the religious issue. However, the high clergy appears to have followed Henry's reforms without too much trouble. One has to remember that there was a significant, caste-like division within the church organization, based on the distinction between

Anglo-Norman settlers and native Irish. While some degree of inter-marriage at most social levels between the two groups occurred, the clergy tended to remain distinct even before the Statutes of Kilkenny (1366) which sought to isolate and retain the superiority of the English over the inferior native population. The main significances of the reigns of both Henry VIII and Elizabeth I, which cover the bulk of the Reformation period up to 1603, were: the reorganization of government in Ireland based on the county system, a significant extension of royal power in Ireland, the successful development of a church hierarchy of reformed bishops, and, finally, a growing popular sense of the inner unity of the island. An island-wide sense of identity does not pre-date this period (Beckett 1969; Boyce 1982: 15–93; MacCurtain 1972).

Even so, Elizabeth's wars wreaked havoc and slaughter upon the inhabitants and left parts of the island in desolation. Attempts by Elizabeth to secure loyalty from the inhabitants by planting protestant settlers largely failed. This is not surprising. The Irish themselves did not recognize a struggle between themselves and an invader. They were used to a state of warring between their semi-feudal lords. Thus fights between a lord and the Elizabethan troops were simply seen as just another struggle among the many (Canny 1982: 93). In addition, these conflicts were in no way perceived as religious, particularly by the native Irish-speaking population. There was simply no knowledge of the Reformation and what was at stake in religious terms. As Canny says, with reference to Irish speakers,

With few exceptions, the bardic poets and annalists, who were not only an educated but a highly privileged order in Gaelic society, were so unaware of what was occurring in England, much less Continental Europe, that they do not seem to have been able to distinguish between the deluge of critical abuse being hurled at their society by New English Protestant aggressors and that which, ever since the time of Giraldus Cambrensis, had been levelled at the Gaelic polity by Catholic spokesmen from the English Pale. (Canny 1982: 93)

This situation was to change from the beginning of the seventeenth century, when young native Irish joined what had by now become the 'Old English'[1] in frequenting seminaries on the continent. These

---

[1] The 'New English' were the settlers and administrators of the Tudor reformed period and of the later Stuart period.

united religious intellectuals took back with them to Ireland the principles and doctrines of the Counter-Reformation to spread among a largely illiterate native Irish population. They did this with the use of the Irish language and produced catechisms, summary treatises, and devotional books. These were distributed to the educated for use by them in the religious education of the populace. Canny notes an emphasis on individual devotion in works directed to the educated, something of a novelty in Irish religion. Hence the spirit of the new catholicism was 'highly pietistic and highly combative'. After the failure of polemical argument with protestants, one could always fall back on the abuse of the reformers:

The horror of Luther's sexual engagements with his wife, a former nun, were described in lurid detail, both Luther and Calvin were charged with being sodomites, and accounts were provided of the sexual misdemeanours supposedly engaged upon by the mother of Martin Luther with the Devil—this last giving origin in the Irish language to the neat patronymic 'Luitéir Mac Lucifer' (Luther the son of Lucifer). (Canny 1982: 98)

The aims of the Counter-Reformation were largely achieved and James I's (1604–23) attempt to anglicize the natives failed. His plantation policy in Ulster—setting up loyal protestants from Britain on land confiscated from the previous landowning élite in the counties west of the river Bann—was much more successful, at least for the portion of territory thus settled. When Cromwell's plantation throughout Ireland (1651–8) was added to that of James in Ulster, the result was highly significant. A powerful protestant presence was established. Their survival and control of the lands, which had been awarded in payment for equipping the army or in payment to the soldiers as a form of arrears, depended on the backing of the English parliament as well as on their own resolve, itself supported by their own religious convictions about their role in the destruction of popery. The Cromwellian plantation reduced catholic landownership from three-fifths to one-fifth of the island.

The protestant planters were largely convinced of the inferiority of the Irish on both religious and racial grounds. They believed themselves to be agents of civilization as well as the true Christians. But they did not form a cohesive and single class interest. Many of Cromwell's ex-soldiers who were new to farming and a number of the financial adventurers who had supported Cromwell's campaigns, sold their land-pledges to fellow protestants. This resulted in the growth of a landed class in opposition to a farm-labouring population among the planters themselves. Also the protestants were divided between the nonconforming groups, largely presbyterian, and the members of the established church. As such, they also had their fights.

The development of presbyterianism in Ireland was preceded by the new order of the Reformation on the English model. Even so, it had a puritan wing, whose principal representative in Ireland was James Ussher, archbishop of Armagh, 1625–56. Though James I of England was also James VI of Scotland and head of the two churches of Scotland and England, these churches were hardly similar. The official Irish church, whose head was also the king, had both presbyterian and episcopalian ministers appointed to parishes. Also it appears that religion was not much of an issue among the planters prior to 1633. But there was a revival movement in the Six Mile Water area covering parts of Counties Antrim and Down. Miller (1978a) speaks of some Scottish settlers as having presbyterian leanings; Barkley (1959) implies that there were many more.

What came to be a significant aspect of presbyterian identity and allegiance to the crown was their covenant, religious–political outlook. This arrived relatively late in Ireland. Scotland in the sixteenth century had, like Ireland, been ravaged by war and unrest and local peoples gathered together to form bands for mutual self-protection. Calvinism developed in opposition to the more Anglican type compromise between reform and catholicism and to this latter's intrinsic support of kingly power in the church as well as in the state. The need to develop a theory of royal power out of their Calvinism to support their vision of true Christianity led the Scots to develop a covenant theology. The cause of the civil war between the king and the Scottish people was Charles I's attempt to impose a sacramental liturgy on the Calvinistic inspired church of that country. The attempt was spearheaded by Archbishop Laud and the Earl of Strafford, who had been the king's representative in Ireland as Lord Wentworth from 1633 to 1640. It stirred opposing Calvinists to put to the king a 'Solemn League and Covenant' in 1638. The document, set forward by the Scottish estates and the Scottish kirk, called on him to accept in Scotland the Calvinist church order and the doctrine of the Scottish kirk. The spirit behind the proposed agreement was the view that the king only had authority over the kirk to implement that which was of divine ordinance. His authority over the state was similarly conceived.

The beginnings of this mode of thought were brought to Ireland by the lowland presbyterian settlers. Its full-blown form can be found replicated much later by the Revd William Trail of Lifford who, in 1861, at his trial before the Irish privy council for opposing the crown, argued: 'I do not believe that the king has power to set up what government he pleases in the Church; but that he has power to set up the due and true government of the Church' (Reid 1853: i. 501; Miller 1978a: 20). The distinction is vital to presbyterian social action in Ireland and had an effect also on the low church tendencies of the established church, from which the presbyterians began to break as Wentworth strengthened royal control over it. But they were to maintain their allegiance to the king even when practically opposing any move on his part which infringed their perception of Christianity—and their perception did affect the political view of society, such as the need for sabbatarian principles to be enforced. When the catholic Irish, led by their partly dispossessed and ever-threatened landed and educated nobles, rebelled 'in no way ... against the king ... but only for the defence and liberty of ourselves and natives of the Irish nation' (Sir Phelim O'Neill's proclamation at Dungannon in 1641, quoted Boyce 1982: 79), the presbyterians of the North banded together after the tradition of the Scots for their own survival. Of course, the rebellion sought at one and the same time the return of confiscated lands and the extirpation of protestantism in Ireland. As a consequence, when presbyterian clergy, as emissaries of the Scottish kirk, came to visit a small number of forts to administer the solemn league and covenant to the soldiers, the ceremony of subscribing to the pledge spread throughout the Northern presbyterian-minded community, with the ministers visiting the people at their own behest. Their banding together had the same sort of self-survival instinct which the political–religious demands of the covenant appeared to express. The presbyterians too were uniting for their own self-defence but they were also doing so in the name of the king. Both the Irish and Scottish presbyterians condemned the English parliament for executing Charles I in 1649.

The Anglican or Church of Ireland laity were already becoming the principal landowning élite and were to become particularly dominant in the southern and eastern provinces of Munster and

Leinster. This factor lead to a radical differentiation between the agrarian economies of the northern province of Ulster and the other two relatively prosperous provinces, Munster and Leinster.[2] In the North, small land-holdings were to become common. The income from these would have to be supplemented, principally by weaving. In Munster and Leinster, large estates were to become the rule, some under grazing and others sublet to numbers of tenants. While economic differentiation thus began to separate North and South, the new, mainly Anglican, landowning élite of both parts shared common sentiments in their awareness of the precariousness of their position in and on a land still claimed by an opposite catholic number.

Anglican dominance was also to be of a religious kind. Wentworth's work in Ireland at least succeeded in establishing firmly the Anglican communion, which would otherwise have disappeared. Its strength lay not only in royal patronage but in the now indigenous population of landowners and landlords subscribing to the established church, together with a significant number of tenants of the same church, scattered throughout Ireland, though with a certain concentration in the southern counties of Ulster. The dual structure of the protestant community in Ireland was established. But protestant Anglicans and the dissenters as a whole, whatever the varying strengths of their religious convictions, were aware that the justification for their presence was both royal or parliamentary decree and the English policy of the extirpation of popery.

The presbyterian church in Ireland grew in part out of persecution. A number of the settlers of the first plantation had left for North America under Wentworth's cleansing of the established church. This reached a climax in 1639 with the requirement of protestants, not catholics, to take 'the Black Oath' renouncing their solemn league and covenant. Despite the restoration of religious freedom in 1641, many more presbyterians fled back to Scotland at the outbreak of the native rebellion. Some of those who remained were slaughtered in the ensuing conflicts (Latimer 1902; Barkley 1959: 9–10; Reid 1853). A Scottish army repelled the rebels from Ulster in 1642, at which time a presbyterian church order was set up for the first time in Ireland, principally in Ulster. The

---

[2] The remaining province of the West, Connaught, was desperately poor, over-populated, and all that was really left to survive on for the vanquished.

Westminster assembly of divines (1642–5), which prepared the *Westminster Confession*, the touchstone of presbyterian faith and polity, established the presbyterian church order for all three countries, and this remained until the restoration of the monarchy in England in 1660.

However, the period was not one of triumph for presbyterians in Ulster. Their view of covenant religious politics was clearly evident in their horror at the execution of Charles I by parliament in 1649. Their refusal to endorse the republicanism of the Puritan party and deny the royal succession led to their persecution by Cromwell's commissioners (1650–4). But with the restoration of the crown in 1660 by presbyterians also came the restoration of the Anglican church order by an Anglican cavalier parliament and the renewed persecution of presbyterians, their exclusion from national and local politics and public office, and the eviction of their ministers from the established church. More of their land found its way into the ownership of establishment supporters. It was then that the presbyterian meeting-houses were built and a clear split between an Anglican and a presbyterian church came about.

For presbyterians, the first mitigation of the harsh consequences of restoration came in 1672 with the *regium donum*, or annual crown donation to presbyterian ministers. It was James II who ended briefly the suppression of catholic and dissenter in 1687; though it soon became clear that dissenters would come off badly should James's plans for the restoration of catholicism come to fruition. This helps one understand the peculiar position of the protestant people of Derry in 1689, claiming James as their king while excluding his army.

The lot of the catholic population and their church was much worse. The extirpation of popery was the aim of both the establishment, low or high church, and of the presbyterian community; though presbyterians had consigned the task, with the *Westminster Confession*, to the 'civil magistrate' or secular arm. The same religious aim was expressed by Cromwell and his planted puritan, low-church ministry during his period of dominance. Priests were persecuted in order to open native minds to the gospel.

However, under Charles II (1660–85), former catholic landowners who had remained monarchists and who had the support of catholic peasants, sought redress from the crown for the loss of their estates and were satisfied in part by partial confiscation of

Cromwellian settlers' lands. So, in addition to the fear which settlers still experienced as a consequence of the slaughter of protestants in 1641, was added the fear of loss of land and livelihood. Matters came to a head with the arrival of the Roman catholic James II on the English throne in 1685. An almost entirely catholic parliament in Dublin then voted itself back all confiscated land. The Act only required a successful Jacobite settlement to succeed. James was also intent on restoring catholicism. Economy and religion went hand in hand. With the banishment of James by the English—to the immense relief of the protestant Irish—the catholic Irish, both landowner and peasant, sought by force of arms to retain James as their own king and thus secure the reconquest of their lands for themselves and their religion. With James at their head, and the assistance of the French, they still lost the war (1689–91) to the protestant settlers and supporting English army, which was led by the new king, William of Orange, consort of Mary, James's protestant daughter.

Very significantly, the presbyterians of the North did not at first take up arms against James's supporters in Ireland, for they accepted James as their legitimate king. Only when James had clearly lost the fight in England and his protestant daughter Mary was proclaimed as successor, did they change their allegiance. In December 1688, the presbyterians and fellow protestants of Derry barricaded themselves in and refused to quarter the local Irish catholic army of James because they could not trust them. They were, yet again, fighting for their survival and were still 'determined to persevere in our duty and loyalty to our Sovereign Lord the King [James II]' (Citizen declaration, quoted Miller 1978a: 23).

The defeat of James finally put an end to catholic fortunes. The settlement that followed resulted by 1707 in catholic landownership in Ireland being reduced from one-fifth to one-seventh and the catholic landowning class, as a power with which to reckon, being destroyed. But, most significantly, it was not King William who destroyed what was left of the catholic landowning class, but the protestant Irish parliament dominated by Church of Ireland membership. According to the provisions of the Treaty of Limerick of 1691, toleration was envisaged and catholics were to remain with the lands they had held under Charles II. It was the Irish parliament, controlled by the landowning élite of the established church, which refused these arrangements.

What emerged from the English revolution of 1689 was a persecuted and poorly organized Roman catholic church, and a persecuted, but well-organized presbyterian church in Ireland. The Church of Ireland was in the driving-seat in both religious and political terms. Its church officials solemnized the marriages of all protestants and looked after matters of education, health, and welfare for the next 150 years. But, in persecuting both catholic and presbyterian, the establishment was exploiting a difference in the material and spiritual interests of the two parties, as Barkley points out:

The argument for toleration . . . lost much of its force because against Romanism the Established Church knew that the Presbyterians had, for their own sakes, to stand by the Established Church in every threat to the Protestant interest. This meant that while the Presbyterians claimed toleration as a right, and the English government urged it as just and expedient, the Irish government, acting largely under the influence of the bishops, refused to grant it as it would undermine the privileges of the ruling class. (1959: 20–1)

In other words, the ability of the establishment or ascendancy to subordinate presbyterians was based on their common interest *vis-à-vis* the catholic population. That common interest was at one and the same time economic, political, and religious. The loss of control of Ireland to the catholic population would have meant the loss both of the wealth, livelihood, religious freedom, and possibly their lives. Also, both groups were aware of the importance of maintaining a stranglehold on the instruments of power in respect of London. The events of the seventeenth century as a whole had taught both protestants, as Anglicans were known, and dissenters, as presbyterians were known, that their dominance and their version of the public order largely depended on their own home-grown resources. London was far away and Irish protestants had learned that the control of Ireland had to be exerted from within if their religious, cultural, and economic power was to be maintained.

THE DEVELOPMENT OF THE ASCENDANCY AND THE PROTESTANT ALLIANCE: THE MANIPULATION OF MATERIAL INTERESTS, 1700–1778

By the end of the seventeenth century, six-sevenths of the land in Ireland was owned by protestants, predominantly in the form of estates and largeholdings. There were a significant number of smallholdings too, predominantly in Ulster and shared by both established church members and dissenters. There was therefore a protestant agrarian tenant class, supplementing their meagre land income from weaving flax. In addition, one large and one small town were in existence in Ulster and colonized by dissenters: Belfast and Derry. In the manners and motivations of the people, it is difficult to distinguish religious concern separately from economic interest and state loyalty. Perhaps the attempt to distinguish between them in retrospect might betray our own cultural tendency to see these states of consciousness and social life as separate rather than separable, because of the ideal rationality of contemporary culture.

Though the need for covenanting by presbyterians was considerably diminished by the decision of crown and parliament to pay presbyterian ministers a stipend through the *regium donum*, presbyterians were not without their internal doctrinal divisions. From 1705, the presbytery of Antrim was ministered by non-subscribers to the Westminster Confession. Non-subscription was based on the belief in the wrongfulness of declaring faith in humanly constructed doctrinal statements as opposed to faith in the word of God, as revealed in sacred scripture. The division was handled in such a way as not to lead to schism. However, by the middle of the eighteenth century, ministers of the separated reformed presbyterians and the seceders, both churches from Scotland, began to develop congregations in Ulster. The former schism had been caused in Scotland by the lack of a covenant agreement on the accession of Anne and William to the throne in 1689. Without such an agreement, some presbyterians refused to accept what they considered to be a degree of contamination within the church of Scotland. The new seceders of the eighteenth century were two separate schismatic groups who broke from the Church of Scotland, the one, known as seceders, for refusing to accept the imposition by the government of a system of minister appointments by local landowners instead of local parishioners; the other, known as the relief presbytery, for refusing to accept any state interference at all in true religion (Bruce 1986*a*). The seceding congregations in Ireland did not accept the *regium donum*, that is until they were offered it—and accepted it—in 1784. But when additional monies were allocated to the mainline presbyteries, the seceders stirred up controversy. Their reputation for greater purity of faith appears to have brought many official presbyterians into their congregations. Later the seceders accepted a further offer of monies from the crown and the ending of this controversy helped towards the union between these churches in 1840. The divisions had been on the basis of maintaining the integrity of presbyterianism and so covenanters and seceders had been anti-liberal and strict in observance. However, the likelihood is that the very plurality of non-conformity in Ulster may have reinforced, for a period, tolerant attitudes among some Northern protestants, as it did in Scotland over a longer period.

Banding continued in modified forms, as landlord and tenant came together to keep public order—that is, the order established whereby protestants owned the land and protestants had priority in working on it. One has to remember that during the same period public order in Britain was achieved by similar means, the local landowner being the magistrate and his own men cast in the role of an agrarian police force.

One can see in these developments the beginnings of the role of protestant–catholic opposition as a key factor in the ongoing production of social divisions in Ireland, to be added to the economic factor of separate agrarian development in Ulster on the one hand and Leinster and Munster on the other. The landowning élite and its clergy of the church of Ireland felt, with some reason, that the dissenters could always be counted on in the event of a catholic threat to the political status quo, as this was so intricately bound up with the joint issues of the dissenters' economic prosperity and religious freedom. The preparedness of the catholic population of Ireland to collude with James, now cast in the role of a pretender to the crown, in his programme to reverse landowner-ship and privilege had its effect in this respect. So, rather than weld into the fabric of dominance the dissenting community, the ascendancy opted for a merely external alliance with them, purely on the basis of two principles: the reformation in the most generic terms, particularly in its political–religious aspect of opposition to catholicism, and anti-native politics. Despite the potential for an absolute and single catholic–protestant divide, other forces were at work to fragment the two main protestant groupings for the duration of the times which counted as normal periods, that is when no threat from the catholic population was apparent.

The new king William tended towards support for dissenters. The Anglo-Irish landowners and clergy exerted their influence to ensure that this support was subordinated to their own interests. They already exerted considerable influence through the Irish parliament. Officially it was subordinated to Westminster as an instrument of government: under Poynings' Law of 1494 all proposals of the Irish parliament required Westminster and royal approval. In fact, until 1719, it had been under royal control rather than that of the English parliament, which had become the British parliament from 1707, by an Act of union with Scotland. This royal dominance had aided the Anglo-Irish purpose. Even after 1719, the Irish parliament was still able to retain considerable autonomy and this was used to further Anglo-Irish interests.

The ascendancy parliament showed its disdain for the dissenters by throwing out Bills of toleration designed to give dissenters legal security, though it appears that the king gave a measure of toleration directly to Huguenots settling in Ireland. These French protestant refugees eventually merged with the presbyterian community. The élite added to their power with the Popery Act of 1704 and subsequent legislation ensuring the disappearance of what was left in Ireland of the catholic gentry. In order to vote, catholics had to abjure. They were not allowed to run schools nor to educate their children abroad. Any remaining control they had of land was undermined by legal changes in the rights of inheritance: an eldest son converting to the Church of Ireland was guaranteed the succession *in toto*, and his father reduced to the status of life-tenant; otherwise, the estate had to be split up among all the children upon the father's death. The wills of catholic landowners were simply declared illegal. No new Roman catholic clergy were allowed to come to Ireland. Those banished who returned were to be put to death.

The establishment sought to inferiorize dissenters as well. To the clauses designed to eliminate catholic power was added a sacramental test. This consisted in the requirement that those seeking election to public or military office must take Anglican communion. The measure was enacted specifically to ensure that the dissenters would be reduced to, or remain in, a semi-subordinate position. It should be noted that, prior to this legislation, the dissenters still had a fair degree of power, particularly in Ulster. Though their landowning interests were probably in decline during this period, they did control the two major cities of Belfast and Londonderry, as Derry was renamed after the siege, and their clergy and local organization had considerable influence over the lay membership of the church, particularly among the presbyterians. In addition, many of the dissenters were wealthy and were behind the development of industry and commerce particularly in the province of Ulster. The two cities of Belfast and Derry were to become the centres of presbyterian economic power as a portion of presbyterians developed into an industrial capitalist class throughout the eighteenth century and on into the nineteenth century. Also, it was the addition of dissenters to the Church of Ireland population which gave protestants a numerical superiority, particularly in East Ulster. But the advantages of numerical superiority were secondary as long as money, the army, and politics could be controlled by the established church membership alone without dissenter participation.

Through the Test Act, the dissenters lost political control of the two Northern cities. Their persecution by the Anglo-Irish establishment was quite severe. While the ministry of catholic priests was recognized, that of dissenters was not: 'the validity of their marriages was denied, they were not allowed to teach in schools, they were compelled to serve as church-wardens, and their right to burial according to the rite of their Church was often denied' (Barkley 1959: 26). Throughout the eighteenth century, over a quarter of a million Ulster presbyterians emigrated to North America to escape the persecution and its accompanying penury, playing their part in the extension of the frontier and taking with them, apparently, a sense of tolerance in religious matters. Yet despite their semi-persecution, those who remained were still prepared to enlist in 1715, when a further Jacobite rising threatened in Britain. A measure of the fluidity of the situation is also indicated by the frequency of what are known as sectarian riots. These were largely among the peasantry and labourers, and were on a confessional basis, between catholic and protestant. Protestant landlords favoured protestant leaseholders and labourers, but were always able to hold out against unrest from this source by offering and sometimes confirming leases to catholics.

One can understand the deep-seated need the Anglo-Irish felt for controlling their own destiny. The English parliament pursued a policy of keeping Ireland poor and non-competitive. The Irish were not permitted to export wool, cattle, or sheep. The prosperity of the dominant class in Ireland rested solely on rent from land. Hence, an antagonism against England and a corresponding desire to obtain as much control as possible over their territory and eventually, if possible, to break the stranglehold of the English government over their confined position, helped weld the ascendancy together in its pursuit of common goals.

In the mid-eighteenth century, the patriot party developed. Its aim was to strengthen the settlers' control of Ireland by striking some sort of pact or contract with the crown and British parliament. They intended to establish a clear constitution for Ireland which would put the running of the island firmly in their hands. Miller (1978*a*) argues that the strategy brought together the traditions of banding and loyalty to the crown, permitting the development of an assumedly agreed theory for a constitution for Ireland. It was hoped that in its turn such an agreement might lead to a sense of political obligation legitimating the use of the band against the English parliament should the position of protestants in Ireland ever be threatened. It was at this time that the ascendancy took on the name 'the Irish people' in its disputes with the English parliament over taxation and policy, similar to the way the colonizers of North America took on the name 'the American people'. But this secular type of contractarian theory was to have only limited success as justification for ascendancy power.

THE INTERNAL THREAT TO THE PROTESTANT ALLIANCE AND THE BEGINNINGS OF CATHOLIC POWER: THE MATERIAL BASIS FOR SECTARIANISM, 1770–1800

During the last third of the eighteenth century the northern agrarian economy began to change. The work of statisticians on the *Statistical Survey of Ireland* around 1800 indicates longer leaseholds than were previously thought to exist throughout Ireland, and very diverse tenant forms in different parts of Ulster. Around Lough Neagh, there was a home-based, capitalist linen-industry, with peasants on six acres employed by master-weavers. In southwest Ulster, commodity farming predominated, with the selling of products to the Lough Neagh and urban area. The population of Donegal survived at subsistence level and County Derry had small-scale commodity farming. The North-East was dominated by larger, twenty- to fifty-acre, commercial units farmed also by tenants (Gibbon 1975: 26). It is important to note that from about 1650 the income from some smallholdings, particularly in the Lough Neagh area, had already been complemented by earnings from the cottage industry of textile production. This industry was relatively successful and developed during the eighteenth century both in terms of market expansion and local investment, as family members began to devote more time to it than to farming. The classic industrialization pattern appears to have occurred first of all in these areas.

Towards the end of the eighteenth century drapers began to emerge as a significant merchant class and these were followed closely by the bleachers who began to buy up cloth themselves and sell the product they finished. The market in England was expanding. Later, in the early nineteenth century, banking was to emerge in Ulster to provide the needed extra capital for expansion. Throughout the period, the increase in the use of machinery, particularly for bleaching, was beginning in Belfast, drawing in the skilled labour from the countryside which could no longer compete in prices and wages with this new small factory system.

In the South, agriculture was still dominant, and was to remain so until well into the nineteenth century. Between the sixteenth and eighteenth centuries, the communal sharing of land by tenants, a characteristic of the pre-Elizabethan period, practically disappeared. Then, from the 1750s, a rapid increase of the population began, possibly as a result of improved living standards brought about by the shift from pasture to tillage and the introduction of the potato (Connell 1950), possibly as a result of the general improvements in health care throughout Europe at the time (Cullen 1972). Many catholics became relatively prosperous as subletters of land owned by the frequently absent ascendancy. A further change was about to happen. These catholic farmers began to enter into the landowning class. Whether this was allowed because of greater tolerance, or whether the tolerance was itself partly the fruit of the increasing financial difficulties of the ascendancy and their desire to improve their profits, has still to be ironed out by historians. Whatever the reasons, it is clear that the rise in population increased the pressure on landholdings. The rise in demand for land led to a reduction in the size of holdings and a subsequent increase in tenant numbers. With rent also increasing in line with demand, the amount of rent collected went up under the double impetus (Crotty 1966).

In the political–religious sphere, a significant development occurred with the change in attitude towards the catholic threat. This had already begun to happen during the Seven Years War, 1756–63, fought between Britain and France. Catholics, including bishops, were beginning to show an acceptance of the status quo in Ireland and a rejection of French interference. Bishops expressed their loyalty to the crown openly in addresses to the Lord-Lieutenant and asked the faithful to pray for a British victory. When the Young Pretender died in France in 1766, an important link with the past was broken, especially as, until that date, it had been his duty or privilege, as descendant of James II, to nominate new catholic bishops in Ireland. In 1774, the Irish bishops sanctioned the new oath of allegiance to the Hanoverian kings of England. The oath involved rejecting papal temporal power throughout the kingdoms, condemning the views that excommunicated kings might be lawfully killed and that heretics were not to be trusted. But despite the increase in practical toleration, there were times when legislation was used discriminatingly against catholics, as in the counties of Cork and Tipperary in the 1760s during agrarian disturbances.

Soon, the Irish parliament began to reverse this institutionalized persecution. In 1778, the Relief Act permitted the purchase of land by catholics, bringing them into the Williamite land settlement and killing off much of the fears of a Jacobite land revolution. The Act was the work of Grattan and the Anglo-Irish patriot party in the Irish parliament, intent on securing passive catholic assistance in the party's battle with Britain for the control of Irish affairs. The sacramental test was repealed in 1780. There was added reason for the ascendancy to repeal it. In 1778, the French entered the American war on the American side. So bereft was Ireland of defence, particularly as it had contributed with manpower to the fighting of the war in the colonies, there was a real fear of a French invasion. A French landing during the previous war with France at Carrickfergus had led to the spontaneous assembly of volunteers to defend Belfast, thus causing the French to withdraw. Despite strong sympathies in Ulster and other parts of Ireland for the colonies' cause, the fear of catholic France was much greater. In the knowledge that they were very much on their own in matters of survival, they revived the banding habit by establishing a more regular militia, the Volunteers. Forty thousand of them were called for and raised on the age-old basis of the landlord–tenant link. These volunteers were therefore protestants, but because they were dual purpose—some to resist the French, others to defend against any catholic threat—catholics in Dublin and Belfast, where radicalism was strong among the Volunteers, were allowed to join despite the law against arming them (Senior 1966: 6–7). In the counties of Antrim and Down, the Volunteers were largely presbyterian.

Hence developments towards greater tolerance have to be taken along with countervailing ones. The emergence of the Volunteers has to be seen also in the context of the Anglo-Irish bid for total power in Ireland. If the Scots had brought to Ireland the practices of banding and the concept of covenanting, the development of ascendancy power brought with it a modified contractarian theory of society tailored to the Irish context and based on the writings of the Anglo-Irishman William Molyneux. His idea that Ireland was a separate society not bound by English law but only by allegiance to the king with whom men of substance had come together in a contract, had a particular affinity to banding and covenanting. It also came to mean something quite specific in the Ulster experience:

Whereas in England the contractarian myth rested upon the social reality of the gentry's special role in maintaining public order . . . and really defined only their political obligation, in Ulster it rested on a different reality. In Ulster each Protestant irrespective of social position was assumed to have a special role in the maintenance of order in a special situation where massive disorder was traditionally expected from one source: the Catholics. The public band was the *ad hoc* community defined by that role and while the community was not a claimant to sovereignty, banding did have the character of a primeval social contract seeking to place its relationship with the sovereign power on a contractual basis or, perhaps more exactly, to get the sovereign power to acknowledge the terms of its trusteeship of authority. The community's essence was that its members could *trust* one another, and no one else. It stood uneasily between those who could never be trusted (the Catholics) and a sovereign power which might be trusted only within limits. The sovereign power—in the seventeenth century the king, in the eighteenth the Westminster regime—should be compelled to agree to very explicit terms. (Miller 1978*a*: 37)

The difference in many others parts of Ireland, including parts of Ulster as we shall soon see, now appeared to be that catholics after all might be trusted within certain limits not to challenge ascendancy supremacy and that Westminster might be now the only real opponent. In 1782, the Anglican ascendancy or 'Irish nation' succeeded in obtaining independent powers for its parliament, often known as Grattan's parliament, after the Earl of Grattan, its principal founder. As such, Ireland became a semi-independent state, keeping the same sovereign as Britain. Though such dominance was only maintained for eighteen years, it is an important historical event. It showed that Ireland could be conceived of as an independent nation-state entity with its own parliamentary structures. The event also reveals how wide a range of social and ideological phenomena the concept of nationhood has to cover, even to the designation of a religious denomination who made up only one-twelfth of the population of Ireland at that time, while the majority remained subordinate to this nation of land-owning families.

In addition, it also shows how it is possible to achieve the trappings and legal frameworks of power at the very point where the class-group as a whole was on the point of steady decline. The decline was to be based on the weakness of its economic power base, rent. The development of the Irish economy as a whole was impeded by the draining of investment from the country by absentee landlords and by the restrictive trade policies imposed from Britain and designed to subordinate the Irish economy to Britain's industrial and trading classes. The only real success was the Irish linen industry based in Ulster, which had been allowed to exploit the colonial market. The élite of this group were the presbyterians and they were now beginning to rival ascendancy power in the North, particularly as the beginnings of a factory system developed in Belfast. But with the coming of the revolutionary war in the American colonies, this trade was in a state of partial collapse. When the war in the American colonies broke out, there was considerable sympathy with them from among the ascendancy, as well as among others.

A further set of countervailing events was the continuance of violence between catholic and protestant peasants. A great upsurge of violence occurred from the 1770s, in the areas of Tyrone, Armagh, and Derry, as ascendancy landlords in Ulster sought to extract massive fines for unpaid rent, and increased rents for the renewal of leases to their protestant peasants. The financial difficulties experienced by tenants appear to have come not so much from pressure on the land, but as a result of fluctuations in the prices of their produce. The subsequent agrarian class movement by these protestant farmers was therefore based on their common instinct for survival and the preservation of their class interests (Gibbon 1975: 29).

Miller (1978*a*: 36) argues that there is little evidence to support the view that catholics suffered in this violence, and that the main object of the movement was opposition to ascendancy landlords. One should point out, however, that it is probably true that catholics did step into such contested leaseholds. One feels it unlikely that they were left untouched especially as such groups as the Hearts of Steel or Steelboys drove out anyone taking up the leases and considered catholics a particular threat:

We are all Protestants and Protestant Dissenters and bear unfeigned loyalty to his present majesty and the Hanoverian succession . . . we, who are groaning under oppression, and have no other possible way of redress, are forced to join ourselves together to resist . . . some of us, refusing to pay the extravagant rent demanded by our landlords have been turned out and our lands given to Papists, who will promise to pay any rent. (Senior 1966: 6)[3]

From 1785—and therefore some time after this outbreak—another type of unrest began to occur. Here, we are following Gibbon (1975), who makes the rigid distinction between unrest among the protestant yeomanry, tenant farmers engaging in some commodity production, and that among the weavers. As weaving expanded in production and manpower, so also did the tendency for landowners to lease at comparatively higher rents, for weavers required smaller plots than commodity farmers. This forced up the price of land for those remaining solely in farming. The consequences were unrest among the yeomanry but an unrest which was directed principally at the landowners.

The weaver unrest came from a different experience. Weaving was not growing solely in terms of workforce. It was steadily converting from an independent occupation to lower proletarian status, as competition from industrial entrepreneurs began to

[3] A Steelboy petition of the 1770s quoted by J. A. Froude, *The English in Ireland*, ii, 121–2.

undercut weavers' livelihoods. The status reduction meant also a loss of income as well as control over production: 'The weaver now found he could be ordered at will between his employers' looms and the fields of some gentleman farmer, to whom he might be contracted out' (Gibbon 1975: 31). This transformation appears to have occurred first of all in North Armagh, the birthplace of Orangeism. However, instead of a situation of class conflict appearing, weavers remained subordinate to their employers and landlords, and the pattern of traditional relationships tended to endure. As landowners in the area grew less powerful economically, the more did their paternal relationships, particularly as magistrates, assume importance. A key status diffentiation in this demonstration of favour, particularly in the light of weaver proletarianization, was that between catholic and protestant. Unrest against catholics appears in this geographical area, where the demand for labourers was beginning to extend beyond protestants to catholics who had previously been subordinated to protestant rights of privilege in the land sector:

Faced with the threat to the principle of homogeneity [of protestant tenant and protestant gentry] which determined the character of the social relations of the community, and its relationship with the world as a whole, the weavers appropriated to themselves the powers of enforcing what they understood to be an integral part of the 1691 settlement. In doing so they were reported to be acting under the belief that they were enforcing the penal laws. (Gibbon 1975: 34)[4]

Catholics, then, were the specific object of weaver wrath. This area of North Armagh is precisely where Orangeism had its origins in the Peep O'Day Boys, who began driving out catholic tenants. Roman catholics made some partial attempt to protect themselves by forming the Defenders. Latimer holds that the confrontations and riots between the two groups were also encouraged by the ascendancy (1902: 177). If the movement is seen precisely as an attempt by the weavers to express their solidarity with the gentry and ensure that their employment as weavers took precedence over catholics, then it becomes understandable why magistrates and landlords were more than lenient with the perpetrators. The new band of Volunteers which was formed in Ulster to quell the unrest

[4] Taken from the famous British government 1835 Select Committee report on Orangeism.

was largely made up of Church of Ireland protestants. It thus engaged in the suppression of catholic rather than protestant bands. By 1791, the Volunteers had secured law and order according to their own interpretation.

The Defenders were a main source of recruitment to the United Irishmen, and a number of them became convinced that the ascendancy were collaborating in the destruction of their homes and livelihoods. In fact the whole process of dealing with unrest among the catholic group seemed to end up being canalized in a general strategy of manipulation by the ascendancy to subordinate to their own cause the protestant working and peasant classes. It makes it easier to understand why the attempt by the English parliament to repeal the sacramental test in Ireland failed in the teeth of Irish landowning and church opposition.

The Volunteers, formed for defence of land and livelihood against the French, had not been financed out of government coffers and remained in existence as local-based paramilitary groups. In Ulster, they would celebrate annually the Boyne victory with marching, particularly in North Armagh. Clearly, such activities were related to the exercise of power within their own domain. But given the prevailing social structure, the marchers appear to have been consciously asserting ascendancy dominance in Ireland over catholics, though unconsciously of themselves too.

However, the Volunteer movement was widespread, inhering in diverse class groups. Thus in Belfast, Antrim, and Down, its ideology was different, showing anti-sectarian tendencies. It is frequently argued that many within the Volunteer movement were more swayed by principles of civil liberty than fear of catholics and that these were mainly of presbyterian background. It might now be useful to remember the presbyterian views on equality in the church together with the fact that most Ulster presbyterians had relatives in the new world who were used to the sort of British interference which had made some of them promoters of liberal and religious-pluralist ideas. Barkley (1959) is in tune with the consensus among the presbyterian intelligentsia in seeing the predominant political attitude among presbyterians of the eighteenth century as radical. Even Gibbon (1975: 34–5) accepts this idea.

Miller (1978a; 1978b) tends to reject this view. On the one hand he identifies the liberal views of some of the ministers, traditionalists as well as the 'new light' party, who can be seen openly to espouse Enlightenment or revolutionary ideas of religious liberty along with a number of educated and prosperous presbyterians. He then identifies the bulk of rural presbyterians, whether involved in linen-making or in agriculture, along with an unspecified number of clergy. The second group were staunch in their support of the original concept of the Calvinistic polity. They believed that only God saved—and probably only a few—but that the obligation to construct a godly civil order still remained. Miller dubs the practical presbyterianism of the Scots and the Ulstermen as 'prophetic social criticism' (1978a: 72), an attitude which was applied to the crown as well as to Rome. While many clergy had been bought off by the *regium donum*, the tenant farmers had remained true to prophetic principles and scorned earthly powers as well as catholicism. Miller points to evidence of millennarian prophecies on the eve of the 1798 rebellion. The presence among the rebels of seceders and some reformed presbyterians, both of which held critical views of earthly power, is indisputable. As for the alleged rationalism of reformed presbyterian rebels who had joined the United Irishmen at the behest of a priest, 'The real key to their behaviour is not reason, but joy: if popish priests themselves were taking a hand in dismantling the outworks of Babylon, the promised deliverance must needs be at hand' (Miller 1978b: 82).

Miller's case is supported by the fact that it was mainly the yeoman presbyterian tenants of Antrim and Down, under threat directly from their protestant landowners, and not the protestant radical weavers of North Armagh who joined the ranks of the United Irishmen in seeking to rid Ireland of both the ascendancy and the British in 1798. In addition, an impartial analysis by Miller (1978b) of lists of presbyterian ministers (30) and probationers (18) implicated in the rebellion shows that half were of radical, 'new light' tendencies and half were orthodox.

If Miller is correct, then rather than see one single uniting cause of disaffection, radicalism, in protestant participation in the rebellion, one is led to identify a multiplicity of conflicting motivations leading different groups to the same course of radical action. One group were the yeomanry, strongly influenced by the prophetic tradition, but having little love for catholics. Second were presbyterian clergy with similar prophetic attitudes and third clergy and laity with enlightened, liberal attitudes. Some of these were among the enlightened Belfast merchants and professionals who ran the radical newspaper the *Northern Star*. The ideas on the extension of the franchise beyond the limitations of religious affiliation could be seen as acceptable in a prophetic context, as well as being supported by the Belfast presbyterians who were indeed radical for totally different reasons.

It seems to have been in these different ways that ideas on extending the franchise began to spread among sections of the Volunteer membership. A new assembly of Volunteer delegates at Dungannon, made up of dissenters in the main, supported this view in 1783 and a Grand National Convention met in Dublin later in the year to promote the position. The Irish parliament, now under full ascendancy dominance, and which had succeeded only the previous year in gaining its independence from the British parliament, threw out the suggested reforms.

A variety of splits cutting across the class and religious structures of the island could now be seen. One may have expected the Roman catholic population to have seen the diversity as an opportunity; but then they themselves may not have been aware of it. In any event, with the coming of revolution in France, majority catholic opinion in Ireland sided against any out and out radicalism. Irish catholicism of the period and particularly of the clergy was strongly influenced by the political passivity of French catholicism, which had entered Ireland via the Irish College in Paris where many home priests were trained (Whyte 1960). Despite this, a considerable number of them still saw advancement for their freedom in an alliance with this rebellious section of dissent. The societies of United Irishmen became the focal point of this gathering of forces opposed to crown and ascendancy rule. Under the radical deist protestant, Wolfe Tone, the two religious groupings, however different internally, found a leader who knew about both sides. He was also secretary of a catholic organization for electoral reform, though, threatened with capture and imprisonment for his radical activities, he went into exile in the United States in 1795.

By 1790, the government in England was terrified that a revolution on the French model might occur both in England and in Ireland, and began to buy off the Roman catholic church and its bourgeoisie. The British government's attempt to introduce a small measure of catholic emancipation were subverted by the ascendancy parliament in Dublin in 1792, though greater freedom of education and admittance to the legal profession were obtained. In its turn, the failure to grant any form of emancipation led to the Catholic Convention of that year, and an island-wide mammoth petition, supported by many presbyterians and taken in procession from Dublin to Belfast and from there to the king in London. The Irish parliament was forced by Pitt and the British government to admit catholics to the suffrage, though, of course, on similar property terms to protestants, with the forty-shilling freeholders obtaining the vote. Pitt was thus forced to modify the royal and class basis of the status quo by admitting such smaller landowners in his attempts to limit support for revolutionary tendencies. Attempts to make further inroads into the ruling class by reforming the boroughs in 1793–4 were defeated. A proposal to permit catholics to enter parliament was quashed by the Irish cabinet, before it even came to parliament. But the crown did obtain a measure through the Irish parliament in 1795 endowing a college at Maynooth as a seminary for the training of Irish priests.

The gains of the catholic gentry and clergy were viewed as dangerous by most Irish protestants. For in addition peasants and labourers had been favoured by the British parliament's need to raise a militia in Ireland as war with France developed in 1793. Catholics were now carrying arms legally and lodges of the Defenders grew up within the militia ranks. These joined the other lodges who had already been involved in activities against both protestant landlords and peasants from 1789, and became increasingly feared by the protestants of Ireland. In addition, the old protestant-dominated Volunteers were disbanded. For Senior, these developments led to the formation of Orangeism:

> These measures dealt a heavy blow to the position of the poorer protestants. Their volunteers were gone, the extension of the vote to the catholics made the protestants less valuable to the landlords, and finally, the militia with Defenders in its ranks seemed to deprive them of the protection of the state. Thus the poorer protestants were thrown back on their own resources to meet the ever bolder and better organized Defenders. (1968: 13)

Gibbon (1975) sees the emergence of Orangeism differently, and coming directly out of the local scene in North Armagh. He notes that the local culture among the weavers was traditionally one of feuding between rival groups and that anything from the traditional pastimes of cock-fighting and horse-racing to fighting could become the site for confrontations between the parties. He points out that it was the proletarianization of weavers and their effort to remain in the protestant alliance that caused catholic–protestant polarization of the feuding tradition. The protestant vengeance groups changed into societies on the model of the freemasons, and eventually directed all their activities against catholics. One such group emerged among the presbyterians, the 'Orange Boys' of a presbyterian farmer James Wilson, in 1793. It is traditionally seen as a consequence of fighting between this group and Roman Catholic Defenders in and around the Diamond, County Armagh in 1795 and culminating in what came to be considered as a battle in September of that year won by united bands of protestants, that the Orange Society was formed. Presbyterian authors traditionally hold that this banding movement was undertaken specifically as a defence against the Defenders (Latimer 1902: 177; Killen 1875: ii. 363). Even so, although Orangeism may have started in North Armagh, one still has to account for its growth elsewhere. It would seem that the very strengthening of Roman catholics as a group was more than enough to do that in a climate of change where the position of protestant groups was under threat.

The bands continued their activities in North Armagh. A number of catholics[5] were forced to flee. Protestant gentry also began to meet to check the Defenders by civil means. Eventually, protestant landlords who saw a common interest with their tenants founded a new society in June 1796, with the intention of absorbing the Orange movement into its ranks, quelling public disorder, and using the resultant lodges as an armed force ambiguously defensive against the French as well as against those threatening their position in Ireland. The ascendancy government colluded in this development of the Orange order. It resulted in a protestant part-time army, on the model of the old English yeomanry. Orangeism now had official approval (Lecky 1912: iii. 421–49).

Probably the participation rate of presbyterians in the order was low outside South Armagh. But there was a significant sharing of views between subordinate protestants both in and outside of the order, whatever their precise class divisions. What sets up a relationship between Orangeism and the spirit of popular presbyterianism

---

[5] Estimates of numbers varied between 180 and 700. See Senior 1966: 30.

for instance, in addition to the defence of protestant liberty, is the tendency within Orangeism to perpetuate the sense of banding, which previously had been the role of the presbyterians, who still retained that tradition within Ireland at least among the seceders. Perhaps the seceders were the only other main group besides South Armagh presbyterians to adhere to Orangeism, despite some of them being United Irishmen. The *Dublin Evening Post*, reporting the famous Orange Day parade of 1796 which virtually clinched the new order's links with the ascendancy, claimed that the participants were 'a motley group of turncoats, Methodists, Seceders and High Churchmen'. Senior argues that the exclusion of mainline presbyterians was deliberate: 'both the *Post and Northern Star* published long articles denying that Presbyterians were responsible for the "Armagh outrages" . . .' (1966: 40–1). Senior could well have been wrong. There does not yet appear to be evidence of mainline lower-class presbyterians adhering to the order anywhere else than in South Armagh.

The general state of unrest in the countryside and the split in loyalties and armed factions should now be clearly apparent. The Irish government took several steps to quell disturbance, permitting some local gentry to arm protestant bands. Now there was a small number of dissenters among the ranks of the United Irishmen and among anti-catholic bands. The presence of both tendencies shows the ambiguity of political forces at that time. The United Irishmen were joined by more Defenders as fear of Orangeism grew. The organization may have been of catholic and presbyterian mix in Belfast, but it was almost entirely presbyterian in the counties of Antrim and Down, and probably made up of many 'prophetic' presbyterians, as Miller suggests. Its rank and file in the South was largely catholic and mainly concentrated in County Wexford. Furthermore, there was a clear sectarianism embedded in the political motivation of many of these catholics. For example,

The oath taken by a new recruit to the united Irishmen in Carlow town in May 1798 bound him, not to 'a brotherhood of affection among Irishmen of every religious persuasion', but instead 'to be true to the Catholic religion and to assist the French should they land in this kingdom'. An almost identical form of words was in use a few months later in Co. Clare. Even in Dublin city the message of the United Irishmen was that 'In a couple of months there will be but one religion, and we will be the owners of the soil.' (Connolly 1982: 9–10)[6]

An added division was between the land interests of the peasantry and the radical politics of the leadership. This spread of interests is important for understanding the course of the failed rebellion.

The leadership decided on armed insurrection in order to set up an independent republican Ireland on the model of the French and American Revolutions. But the rising of 1798 was badly organized and the leadership was arrested before the men took to the streets. The imbalance of aims which came to light as a consequence added to the failure of the movement. Catholic peasants in the Wexford area, South-East Ireland, went about settling old scores, killing or maiming protestant landlords, and a number of protestants were burned to death in a barn at Scullabogue. As the rebellion in the South started a fortnight before that in the North, news of the sectarian character of the Southern rising led to many Northerners defecting from the ranks. Tone, who had sought the aid of the French government, joined one of two small invading forces. While the smaller one landed in Ireland and had some initial successes before capitulating to a much larger British force, the other never landed and was put to flight, Tone himself being captured (Lecky 1912: iii. 504–39). Protestant peasants and workers were now left in no doubt as to where their true interests lay, namely with their protestant overlords. In Ulster, two presbyterian clergymen were executed for their part in the rebellion, a number of others were imprisoned and the remainder emigrated. The synod of Ulster disapproved of rebellion and sought constitutional reforms.

Thus the rebels' official aim to bring about a new kind of Irish nation was defeated, and the opinion that catholics were subversive and traitorous reconfirmed in the minds of many protestants. The failure of the rebellion meant the end of Ulster dissenter support for an independent, multi-denominational Irish nation, independent from Britain. The contract which had appeared to include catholics could now no longer be made, and the following century is the history of protestant Ulster's attempt to clarify what this might now mean.

---

[6] How the last sentiment from Dublin is to be reconciled with the local leadership's supposedly non-sectarian views is not clarified by Connolly. The reference is from *Rebellion Papers*, State Paper Office, Dublin.

The very fact that rebellion had occurred also spelt the beginning of the end of ascendancy dominance. The Anglo-Irish had showed themselves unable to deliver Ireland into Britain's imperial pocket. So Pitt dissolved the Irish parliament and merged Irish representation into the British parliament by the Act of Union of 1800, practically dissolving the protestant Irish nation in the process.

3
Religion and the Growth of Opposing Power Groups
Ireland on the Road to Division, 1800–1922

THE DEVELOPMENT OF PROTESTANT–LOYALIST POWER, 1800–1905

Introduction

This chapter reviews the decline of ascendancy power and the emergence of the two opposing alliances which have dominated Ireland's internal relationships to this day and which lie at the root of the troubles. From the beginning of the nineteenth century, the ascendancy began to lose land to catholic tenants. The Act of Union of 1800 which had abolished their Irish parliament reduced their political power at Westminster by abolishing the rotten boroughs.[1] The ascendancy was made to struggle to regain influence with the odds stacked against them. The succession of reform Acts enfranchising presbyterians and catholics added to their decline. The last vestiges of their direct religious dominance came in 1869 with the disestablishment of the Church of Ireland. Their economic dominance was finally ended by a series of land Acts between 1870 and 1909 which gradually transferred ownership of land to their tenants. Only ascendancy landowners and gentry in the North East were able to turn elsewhere to retain any real power. A new form of protestant alliance had already started throughout Ireland in 1795 with the emergence of Orangeism but was of minor proportions. The split between Northern and Southern ascendancy landlords only occurred about 1900, when the impossibility of ever regaining control of Ireland was finally recognized and Northerners went their own separate way. Northern ascendancy landowners, and smallholders and tenants of both protestant traditions then acceded to a new joint alliance with the Northern industrial élite and became incorporated in the protestant–loyalist alliance of Ulster.

[1] Rotten boroughs were parliamentary constituencies with no voters, but owned by members of the aristocracy or other powerful figures. The owners simply appointed themselves as MPs or gave the seats in patronage.

The conditions for the development of an Ulster protestant centre of power already existed at the beginning of the period. There was numerical superiority in half of the province and a growing industrial and financial bourgeoisie. Dissenters led this modern economic group but they were not yet well disposed to the ascendancy. They were, however, soon to become strong enough to be able to forge a durable alliance with them. In addition, there were protestant workers and peasants who were reliant upon the upper classes for their livelihood and patronage over against the catholics.

The possibilities of a new centre of protestant power were far from clear in the early half of the century. This lack of clarity was shown in the changing fortunes of the Orange order. At first it drew on the membership of the established church throughout Ireland, as well as on presbyterian weavers in North Armagh, incorporating peasant, worker, and employer alike. With its meetings, inner secret societies, and pledges both to the religious principles of the reformation and to the crown, its organization and ideology emerged strengthened by the experience of the 1798 rebellion. The commitment required of its members was explicitly both religious and political, at a time when political institutions throughout Europe were becoming more differentiated from the churches. In the early years, the order had taken contractarian politics seriously. There was a conditional oath to support the crown 'so long as he or they support the Protestant ascendancy, the constitution and laws of these kingdoms' (Miller 1978: 61). In 1823, the British government succeeded in suppressing the oath, though it appeared again in evidence taken by the Select Committee of 1835 set up to investigate the order's activities.

The order was claimed to have a membership approaching 200,000 in 1935 by one of the order's spokesmen, who gave evidence to the *Select Committee* of 1835. However, by 1851 there were only 1,335 lodge members in Belfast at a time when Miller argues there must have been about 90,000 in Ireland as a whole (1978: 58). Orangeism thus declined significantly in the period. Its decline was partly brought about by Drummond, the Under-Secretary for Ireland, who attempted to suppress it in the late 1830s. Still, the order managed to survive. By 1870, it had reached in Ulster its lowest point and consisted, in the countryside, almost entirely of the alliance between church of Ireland landlord and tenant, with an unspecified membership in Belfast. From then onwards, it grew again, under the impact of the catholic–nationalist threat, to play a significant part in the formation of the protestant–loyalist alliance.

Presbyterian participation in the order does appear to have been low until the 1880s. Antagonism between dissenters and the ascendancy was probably still a strong element in Ulster politics, though Miller argues that the slow growth of Orangeism in Belfast was due to 'the absence of a readily visible Protestant élite anxious to act out the special relationship' with protestant workers (1978: 58). But the slowness of presbyterians of all classes to cement any alliance with Anglicans might provide just as convincing an explanation as the one Miller provides.

Different groups of presbyterians appear to have followed different politics according to their interests. Presbyterian farm-tenants were involved in the battle for land reform and had alliances with catholics. There was a similar alliance between presbyterians and catholics in the South before the activities of the more nationalist oriented Land League began in 1879. The duality of opposition between ascendancy and dissenter, and dissenter and catholic was clearly continuing. These antagonisms, coupled with the diversity of political–religious motivation between and within protestant groupings, rendered any developing protestant alliance volatile. It may have been that some alternative form of alliance, a non-sectarian one based on liberal or radical values, could have grown in the decades around 1800. The presbyterian radicals of Belfast and Dublin had a real sense of tolerance towards catholics, and this sentiment may have been shared by a number of the weavers of Ulster.[2] Also, within the collective beliefs of presbyterians, there was also a long-standing sense of protestant liberty, going back to the *Westminster Confession*. This belief is still preserved today and proclaimed at the ordination of presbyterian ministers and church elders, and at the installation of ministers:

God alone is Lord of the conscience, and has left it free from the doctrines and commandments of men which are in anything contrary to his word, or

---

[2] Gibbon 1975 uses information on the rhyming weavers of Ulster and the known radical literature of their local libraries to support the view that such a radicalism was widespread in Ulster as a whole among presbyterians. But these semi-independent weavers were declining in numbers by this time. The idea of widespread tolerance among presbyterians is questionable.

beside in matters of faith or worship. So that to believe such doctrines, or to obey such commandments out of conscience, is to betray true liberty of conscience; and the requiring of an implicit faith, and an absolute and blind obedience is to destroy liberty of conscience and reason also. (*Westminster Confession*, Ch. 20, para. 2, quoted Thompson 1981: 15).

On the one hand, the statement is directed against two aspects of catholic discipline: the demand for obedience on all matters declared to be of faith and morals, and the demand for an implicit faith in the teaching of the church even on matters of which one is ignorant. On the other, the statement includes toleration of dissent. In many ways catholics could be seen to be in the same boat as presbyterians under the ascendancy. Perhaps this sentiment could survive only as long as catholics remained either a subordinate or equal group, but not a dominant one able to impose its beliefs of social order upon others.

Further indication of liberal tendencies among presbyterians in Belfast was their support for the liberal party as soon as certain classes among them acquired the vote in the 1832 Reform Act. Such support was to remain steadfast up to the first moves for home rule in the 1870s. However, the entry of working-class groups into the franchise in 1867 was destined to weaken the liberal party. Working-class support for the Orange and protestant Irish cause was to develop further. It is likely that for the bulk of presbyterians, the religious antagonism dating back to the reformation was deeply felt throughout the period. Traditional fears of the catholic threat were firmed up by the events of 1798 and remained bedded in the popular psyche throughout the next century.

Preparing the Way for Dominance: The Role of Popular Protestant Religion in the Industrialization Process

What practical tolerance there had been declined over the course of the century. The growth of the catholic–nationalist tradition played its part in this decline, taking catholics out of the boundaries of tolerance, and aiding the growth of stern opposition on the part of the Northern protestant communities as a whole. Orangeism emerged as the prime institutional focus of such opposition. Though it later came to have a secondary role to the Unionist Party which became the centrally organizing focus of protestant–loyalist power, the order was primarily responsible for obtaining the local solidarity of protestant–loyalist communities. It also fostered the dominant culture which was to provide the basis for unionist ideology as a whole. Though Gibbon (1975) asserts that the order only became a significant force in unionism in 1911, it has to be remembered that much of the foundation on which unionism built had been provided by the order.

But opposition to catholicism and the existence of a local club for protestant meetings and parades were hardly sufficient to construct a resourceful political philosophy able to weld together disparate social groups and maintain dominance over their political thought and action. Further elements were necessary and some of these emerged from religious and class developments within Ulster during the nineteenth century. Miller (1978*a*; 1978*b*) gathers his reflections on religious change in Ulster in the context of Gellner's theory of modernization. Gellner (1964), in this early work, argues the need for a nationalist belief system if groups and individuals are to handle the transformation from a traditional to a modern society. He grounds the consequent development of nationalism in a theory of uneven development. People who have left the ancient world of stable relationships behind require an identity capable of coping with the varied and changing roles one has to play in a modernized society, where belief is more important than in a tribal one. In tribal and peasant societies, beliefs are not so immediately and strictly necessary as people find themselves always performing the same tasks within the same community. The process of industrialization affects rural as well as urban settings. It is not only the people who move to the city for work who are affected. The owners and workers of the land experience dramatic changes. Land shifts, in thought as well as in deed, from a repository of traditional values to a commodity, to be bought and sold on the market. Loyal, semi-feudal workers become part of the cost factor of the farming operation. Nationalism provides a way of immediately establishing significance for one's varied and changing roles by the very fact of being a member of the nation. The changes are consequent on the loss of peasant holdings and the movement from self-sufficiency to a wage economy. The changing roles may lead to several uprootings of household and family in order to follow the labour market. National identity, reinforced by the degree of literacy and education which modernization requires, becomes all the more important when there are labouring newcomers seeking also to enter the world of affluence and competing with an already established proletariat. The use of some label other than skill to exclude the newcomers, such as colour or religion, can become in its turn the spur of some new nationalist form among the rejected.

Transferred to the Ulster context, Gellner's theory is partly convincing. The population of Belfast grew from 13,000 in 1782, to about 50,000 in 1831 and to 350,000 by 1900. As important was the fact that large numbers of catholics entered the city, changing their percentage distribution in the population from around 8 per cent to 33 per cent over the same period. Church of Ireland weavers also moved into Belfast, likewise helping reduce the previously dominant presbyterian element in the town to a third of the population. Probably they also brought the Orange lodge system with them into the city. Such an event could by itself account for what growth in the order there was in the city in the first half of the nineteenth century.

As already mentioned, originally part of the ritual of the Orange order had a conditional oath to the crown, one which reflected the age-old tradition of banding and covenant politics. In 1823, the oath was made illegal, along with the secret oaths of other societies. In any case, since 1800, upper middle-class protestants had found the conditional oath an embarrassment. Their political rights had been achieved and the danger of republicanism had been averted, so they had no further use for covenant politics. According to Miller (1978*b*), this class appears to have claimed in a retrospective way that the lower classes had been instrumental in the inclusion of the oath in the first place. Miller grasps something of considerable importance here. The protestant gentry were now able to identify with the British nation, as their class and status position was assured. But the lower classes were still left with banding as a source of their security; and it had to be from this that an ideology suited to the new capitalist structure had to be born. In a sense, the original purpose of the Orange order was being continued, but a contradiction within that purpose, and which had been implicit in the duality of class composition, was now being exposed at the ideological level. In the 1880s, the subordinate protestants were to show themselves loathe to give up their commitment to self-defence which Gladstone's liberals were seeking to wrest from them by setting up an impartial police force to enforce equality between catholic and protestant. Shortly after, with majority catholic rule looming, the upper classes themselves were to return to the banding ideology. A further item to remember is that such developments must have put considerable pressure on the majority of presbyterians who still remained outside the Orange order. Their position was bound to be re-evaluated as power moved in the direction of catholic nationalists.

The political role of evangelical religion

The protestant revival of 1859 provided a partial substitute for nationalism in mediating modernity to Ulster's new working classes, still bereft of any clear national identity. It has been argued that the revival replaced intellectual presbyterianism with salvationist enthusiasm and left many in the working class more open to enthusiastic Orangeism and unionism (Gibbon 1975). The weakness of this argument is that traditional presbyterianism was less intellectual than is frequently supposed. Also, revivalism was hardly new. Apart from evidence of its existence in the seventeenth century, there is clear evidence of its continuation and growth in the evangelical crusade undertaken by Wesley and others in the latter half of the eighteenth century. From the 1740s, Wesley himself propagated the message in Ireland and had a significant effect on both the Church of Ireland and the presbyterians. The founding of the Evangelical Society of Ulster took place in the same year as the united Irish rebellion of 1798 and both presbyterian and Church of Ireland people were members (Miller 1978*b*: 85). Some commentators have argued, in line with the Halévy thesis, that the revival helped the suppression of the traditionally autonomous and aggressive presbyterianism which had appeared in the rebellion. The movement certainly occurred at a time of industrial slump and may well have aided the successful attempt of factory owners to reduce wages while at the same time developing greater bonds of loyalty with their protestant work force (O'Connor-Lysaght 1980). One can also add that the protestant working class were expected to follow the religious and moral culture of their protestant masters: thrift, industry, sobriety, and clean living. In the Ulster context, however, the pact would have been all the more strengthened if such a protestant ethic were reinforced by a political–religious basis common to all the denominations. It appears that revivalism was able to do precisely this. Aspects of revivalism were not outside the common protestant heritage. Perhaps a novelty of the revival of

1859 was not so much that it was multi-denominational within protestantism and had its strongest appeal to the working classes, but that it was highly enthusiastic involving prostrations, uncontrollable displays of movement, even hysteria and comas, and centred specifically on the felt salvation experience, betraying its origins in North American revivalism of the day.

Throughout the period, the conditions for the appeal of revivalism to protestant subordinate classes of all church persuasions were precisely those of modernization. As Miller argues, the traditional prophetic presbyterianism could only survive if there was some real possibility of a presbyterian polity: a society in which godliness could be enforced by the civil power. In fact, it could only survive in a pre-modern society, one where issues of salvation could be subordinated to the ever-possible intervention of God in history. Weber has argued that there is an essentially rationalizing tendency to the capitalist industrial process which makes belief in the religious relevance of such a world very difficult to maintain (1930: ch. 5). God easily becomes relegated to the realm of individual experience. Conversionism is the pre-eminent form of such a privatized religious experience, and helps explain both the growth of methodism within the Church of Ireland, and the transfer of presbyterian attention away from matters of polity to those of personal religious experience.

But conversionism was not simply politically passive, merely leading to the abandonment of the presbyterian polity, but had an active dimension—proselytism. Before the arrival of conversionism, the three main churches had directed their religious pastoral work solely to their adherents. Now, converts to evangelicalism in the two main protestant churches addressed their preaching to each other's flocks, the sects, and catholics. From the 1820s, proselytism came to be the purpose of a new organization known as 'the Second Reformation'. The Revd Henry Cooke, a supporter of orthodoxy in presbyterian matters,[3] was also an evangelical and played a significant part in bringing presbyterians of this mould together with Church of Ireland members of like inclinations. The movement led to a swing towards enthusiasm and away from intellectualism within organized presbyterianism. Probably, the shift was

---

[3] Cooke played a principal part in the so-called second Subscription Controversy of 1826–35 by which the unitarian tendency were forced out of the Ulster Synod.

gradual: converted elders would have seen to it that the new ministers they appointed to their own local congregations would have had the seal of the conversion experience. Over a period of time, this would have sufficed to modify the balance of power in the general assembly.

To this second reformation the catholic clergy reacted vigorously, as will be seen. But the political thrust of the movement went beyond proselytism. For it was this form of evangelism which led to riots in 1857. It is possible that the street violence itself was related to a shift in ethnic consciousness the new evangelism brought about. The whole thrust of the salvation experience was that it happened to those who had the pure Gospel announced to them. Catholics came to be experienced unambiguously for the first time as the unsaved by definition, for the damned could now be pointed out as those who did not proclaim the felt experience of salvation (Miller 1978; Hempton 1980).

Thus evangelism provided common ground between members of protestant churches and sects, an internal form of settlement between previously sparring parties. What the revival of 1859 appears to have done was to extend the attitudes of missioning evangelicals to many among the new working classes of Belfast and to other uprooted groups who had previously been untouched to any significant degree by such movements. In this way evangelical religion was performing the role of Gellner's nationalism, as Miller suggests.

There must surely have been some connection between the growth of enthusiasm and the growth of the Orange order. Both were rooted in similar social experiences: insecurity over jobs, the processes of proletarianization, the frustrations of poverty, deferred marriage, and family break-up. As the catholic–national movement grew there was additional insecurity over the future of the protestant population in Ireland. If theology had by now become a largely theoretical issue within other reformed states, it was to remain one of political import in Ireland. Anti-catholic theology had an elective affinity[4] with the catholic–nationalist threat of political oppression. The achievement of material and political supremacy by the ascendancy coincided with the believed

---

[4] The term 'elective affinity' is Max Weber's. He uses it to describe the close relationship a group's ideas or beliefs can have with its material interests. See, for example, Gerth and Mills 1946: 284–5.

achievement of moral and spiritual supremacy. The tactics to achieve both were convergent. The threat of catholic–nationalist domination supplied sufficiently added motivation to urge on the process of protestant–loyalist group formation in a similar way.

Value Change among the Belfast Protestant Workers

The development of these specific political–religious attitudes was complemented by changes in other aspects of popular belief, particularly among protestant workers in Belfast. These changes have been pointed out by Gibbon (1975) in his study of working-class protestant consciousness over the crucial years of 1850 to 1886. He shows how values further conducive to protestant loyalism developed out of the ghetto experience.[5] He identifies two types of riot corresponding to two types of protestant working-class experience. The early type is the rioting which occurred at the boundaries between protestant and catholic working-class areas, as in the riot between the protestant Sandy Row and catholic Pound, later to be known as Lower Falls, in 1857 and 1864. Gibbon points out that these riots occurred annually around 'the Glorious Twelfth', were limited to the boundary areas where they occurred, and were begun by groups of boys and youths, who were then joined by young working women. The states of consciousness displayed in these encounters were perceptions of strict, unequivocal segregation of the areas, the embattled experience of a transgressed community, and the sense of restoration on the successful completion of the riot, which often involved evicting tenants who had crossed the line demarcated by the offended group. The moral virtues held in high esteem were loyalty to the local protestant quarter, and being stout, bold, and redoubtable in its defence. The communities in which this ideological form predominated, as in Sandy Row, were communities of unskilled workers in the linen trade, often living in the quarter built by their own mill owner or employer, who frequently drew labourers from his own religious grouping and certainly avoided free labour markets. The workers tended to be primarily women, and basically were the industrial proletariat of the first stage of industrial revolution. They held to local traditional values, rather than to an

[5] Gibbon develops his thesis in total opposition to Boyd (1969) who considers that there is a basic continuity in the nature and structure of Belfast rioting from the 1850s through to the 1880s.

identity based on any specific role in the industrial process as such.

The later type of rioting, typified by the Shankill riot of 1886, had a different set of characteristics. Such riots spread throughout an area and spilled into others also. They were spearheaded by an *occupational* group, the Belfast shipworkers, the élite of the working classes who were both skilled and relatively wealthy. These workers recognized their own clearly identified interest in capitalism and British imperialism. Their very livelihood depended upon the development of empire, trade, and the demand for ships. When the 'Islandmen', as they were known, started a riot, they called on their occupational relatives, the apprentices, to assist them. Then others, mainly young people and women, might also have got involved. Some of the moral perceptions of the shipworkers were shared by the earlier rioters: the sense of embattlement and the need to separate out and restore the established order of superiority and separateness. But the ethos was otherwise different, cosmopolitan as opposed to localized, industrial–occupational as opposed to traditional–personal. In addition, there was a sense of the occupational group as overseers of the progressing battle, of those who must be vigilant in the discharge of their responsibility for the protection of the entire protestant community.

Though the difference in the rioting is remarkable in its indication of a shift in consciousness towards unifying protestant values, there remain the elements of commonality between the two types of rioting. They were both sectarian and directed specifically against catholics. The 1886 riot represents a more advanced industrial and capitalist phase, where catholics become identifiable as an economic threat to the protestant working-class hold on strategic occupations in the industrial labour market. This generalized working-class antagonism has to be accounted for eventually in a historical assessment of the growth of the two opposing interest groups in Ireland. Clearly these elements within the new protestant working-class consciousness were grafted on to a popular religion which pre-dated the generalized protestant identity of the Islandmen, a popular religion which went on to enter the general culture of the protestant–loyalist alliance.

The occasion for the riot of 1886 had been the expulsion from the docks of a group of catholic site-builders, threatening that they would soon have all protestants off the docks. The sentiment which these probably transient catholic workers expressed had a continuity with traditional as well as modern universalistic ideology. It articulated themes of 'who Ireland belongs to' and 'who are in the right', in terms of both justice and religion.

One also has to remember that the riots of 1886 occurred after Gladstone and the liberals had committed themselves to home rule for Ireland and that a real fear of catholic domination, embracing but surpassing fears of employment status, underlay public disorder. The protestant working-class leadership was already locked into a fight for solidarity with the new protestant leadership of Ulster, whatever else happened in the rest of Ireland. Along with other groups of protestants in Ulster, with interests varying from militant evangelism to British imperialism, the working-class leadership needed an alliance against the catholic–nationalist threat. Values and beliefs which could solidify such an alliance were bound to be in the ascendant, and the religious anti-Romanism and cosmopolitanism of evangelical revival, together with the unifying working-class values of the protestant shipworkers, were clearly fitting the bill.

Clinching the Alliance: The Political Drama, 1884–1905

So far, the only pan-Irish institution providing the protestant groups with an organization promising to deliver security on the basis of a set of common deep-seated beliefs and values had been the Orange lodges. Their self-justification was the preservation of protestantism as both a religious and political bulwark against Irish catholics. In addition, each separate lodge was grafted into the traditional pattern of rural relations, without much power being exerted across the order's island-wide organization. Local, personalized patronage as opposed to occupational patronage was its strength. But two external events were to change all that. The first was the major extension of the franchise to all male householders by Act of parliament in 1884, followed by the enforcing of secret balloting four years later. This reform revolutionized the structure of institutionalized politics in Ireland. The numerical majority of Roman catholics in the island now became a political majority. With the right alliance in Britain, catholic domination in the island could now be legally achieved. The second was the conversion of Gladstone, the British liberal prime minister, to home rule, and his immediate pursuit of it as a major political goal. If any representation of non-nationalist, pro-protestant interests in Ireland was to be organized, a rapid transformation in the existing divisions of protestants in Ulster had to take place. As indicated, there was already sufficient basis for this to occur. However, the first attempts to pull the various interests together in the 1885–6 period were poor. Protestant unity was so weak that, had Gladstone's liberal party not split in 1886 over the issue of home rule, an Irish parliament might well have become a fact and could have pre-empted the formation of the protestant–loyalist alliance (Gibbon 1975: 130). Even so, protestant–loyalists did move quickly in the direction of unity. Already, in 1886, protestant clergy were made honorary members of the Orange order, and chaplains participated in varying degrees in the activities of the lodges. The presbyterian bourgeoisie abandoned the liberal party and joined the new unionist movement, even though this meant a degree of subordination to the majority of former conservatives. At this point the landed aristocracy were practically outside the alliance and only came round to it gradually over the next few years.

When the next home rule move was made in 1892, the bourgeois protestants of Belfast took the lead and by-passed the Orange order. They were sufficiently strong to set up the famous Unionist Convention of 1892, which was efficiently organized on the basis of the urban and rural professional classes. Their central political ideology was what Gibbon identifies as social imperialism. The call to organized action was based on what would happen to protestant jobs and businesses if the imperial connection, on which all the prosperity of Ulster was based, was removed. Again the crisis passed, with the second failure of a home rule Bill in the English parliament. When a third crisis came in 1905 and even the British conservative party seemed intent on some measure of devolution, the Orange order was incorporated into the unionist organization. It was given 25 per cent representation on the newly formed Ulster unionist council and other committees. The urban bourgeoisie and workers constructed a joint industrial policy. Leadership was at least officially by the rural aristocracy of Ulster, and protestant church leaders gave active support, as did the faction of the British conservative party which was led by Lord Randolph Churchill.

The previous organization of 1892 had failed to endure. The movement had relied on the Ulster unionist clubs to deliver the protestant working classes to the movement. But the clubs had withered away once their specific purpose of existence had disappeared. The sporadic threat of home rule by itself appeared insufficient to give the alliance stability. The Orange order was called upon seemingly because its more enduring virtues were recognized. It was already in existence as a locally binding protestant force, both in town and country. Also, in addition to the ritual marching on the anniversaries of the Boyne and Londonderry, other forms of communal liturgical practice had developed as part of the permanent initiation into secret society membership. The binding together in consciousness of specifically protestant themes with past events and present problems, was done through society rituals and the very act of meeting for talks and regular sermons. The public marches celebrating the events of the Williamite revolution had never disappeared and in their celebration of protestant history, they carried their past with them into the present conflict, now functioning to mark out protestant territory which would never be surrendered. It was vital that those past events should be seen and confirmed as the decisive victory over the former inhabitants of the now protestant lands. The recovery and development of Orange power adds strength to Miller's contention that the order constituted a revision of the public band, but placed on a surer, institutionalized footing, and renewing in a firmer and more status-conscious way the links between the protestant dominant groups and the subordinate protestant professional workers, the unskilled, and the land labourers and leaseholders.

The political development of a popular protestantism which had on its inside a positive anti-catholicism is clear. It was certainly strengthened by the revival of 1859. The only underlying continuity in its ideology which can be singled out is reformation religion, which in Ireland tended to move around Calvinism rather than Lutheranism. Members of all protestant and Anglican churches tended to remain low-church and evangelical when episcopalian, and either polity-oriented or evangelical when presbyterian, but always anti-catholic in both churches. Theological rhetoric made clear the message of this religion, so that popular anti-catholicism as a religious experience thrived whenever there was rivalry over land and labour between catholic and protestant and whenever there were threats to overall protestant dominance: the rebellion of 1641, war of 1689–91, rebellion of 1798, and the home rule movements. It thrived also in the context of the land wars and urban proletarian rivalry. The Orange order provided the institutional basis for this popular, pan-protestant religion, even though its membership did not extend to all protestants. The order's ability to rise to the challenge after 1905, and the fact that it was called on by the urban bourgeoisie in the first place in order to solidify the Ulster alliance, is indicative of its cultural power. Leading presbyterians had entered the alliance in 1892 and soon went on to membership of the Orange order as well, ending permanently the order's nineteenth-century Anglo-Irish character. The recall of the order welded popular protestant anti-catholicism to the operations and ideology of a class alliance in Ulster, whatever the public, external face protestant loyalism had to Britain and the rest of the world. Religion within the protestant–loyalist alliance was a key force and remains so, as will be seen.

THE DEVELOPMENT OF CATHOLIC–NATIONALIST POWER: THE ROLE OF POPULAR RELIGION AND THE CHURCH, 1800–1922

Forms of Nationalism in Ireland

Over 800 years of Irish history, Boyce (1982) identifies no fewer than three strands of nationalist sentiment inhering in three distinct social groups prior to the nationalism of catholic nationalists. The first were the Gaels. Their indigenous culture was drawn on by the bards to build up the consciousness of a more unified ethnic group in the face of colonization between the twelfth and sixteenth centuries. This has to be considered a very weak form of nationalism, if indeed one wants to call it nationalism at all. The second group identified by Boyce were the old English, the descendants of the Norman colonizers of the twelfth century, who were the first to develop the concept of the 'Irish nation'. Thirdly, there were the sixteenth- and seventeenth-century landowning settlers and members of the Church of Ireland, together with old English converts to the Anglican communion, who, as has been seen, appropriated the concept of the Irish nation as 'the Irish protestant nation'.

The definitions of these groups varied according to circumstances. In the Norman period, the several Gaelic chiefdoms and clans, in themselves distinct, viewed themselves as a group in opposition to the Norman Anglo-Irish. But against the English government of the time, the two groups had some sort of loose alliance, while retaining their distinct identity. Gaels were considered ethnically inferior and the old English even had a separate church organization on near racialist lines. It was only in the early sixteenth century that Gaelic chieftains were allowed to sit in the Irish parliament.[6] The Elizabethan and Cromwellian plantations led to the demise of the power of the old English who eventually merged with the rest of the inferiorized population.

Nationalism has a varied content of beliefs throughout the world. In the Irish context, religion developed a close association with nationalism only in the nineteenth century. Explaining the growing dominance of Irish catholics in this new form of nationalism in Ireland is partly the purpose of this chapter. As will be seen, after the failure of the 1798 rebellion, the movement for an increase in power on the part of the subordinated catholic population took on a reformist purpose. Under the leadership of Daniel O'Connell, violence was rejected. However, from the beginning, it was a movement in which protestants did not really find a place. The anti-Ascendancy movement had lost its interdenominational character. From then on, protestant participation in movements for radical change was only by minority groups of intellectuals, and usually only when it was inspired by separatism.

Economic and Demographic Transformation: Growth of a Catholic Landowning Class, Death of the Destitute

As can be seen in Table 3.1, by the early nineteenth century catholics evidenced a range of class interests. There were cottiers, labourers, and smallholders, all bound into a complex set of independent relationships with the landowners and their managers. Then there were the farmer-landowning classes, whose composition was to change quite significantly already by the 1840s. The penurial state of half the population indicated not only the everpresent threat of famine but also its total dependency on the landowning population and their administrators. Catholic industrial interests only really emerged in the twentieth century. In the nineteenth century, agrarian and commercial concerns were at

---

[6] Historians still appear to argue about the degree of distinction between Gaelic and old English lords, while allowing for the 'gaelicization' of a minority of the colonizers.

TABLE 3.1. Social Structure of Rural Ireland, c.1841–1845

Status	Holding (acres)	No.	Total
Farmers	Above 50	70,000	
	15–50	207,000	453,000
	15 or less	176,000	
Smallholders	Above 5	135,000	
	1–5	182,000	408,000
	1 or less	65,000	
	Unclassified joint tenancies	26,000	
Cottiers and Labourers	N.A.		596,000
Farm Servants	N.A.		104,000

Note: 'Cottiers' refers only to those receiving their wages in land.
Source: Connolly 1982: 17.

the centre of economic activity in the South. The defeat of the united Irishmen in no way prevented catholic middle-class agitation for greater economic and political power within the prevailing British regime. The bulk of them had been opposed to revolution. As the movement for catholic emancipation under O'Connell developed, it was the interests of catholic farmers buying ascendancy land which came to the fore. On their behalf Daniel O'Connell formulated his policies and developed the Catholic Association.

During the century, the land question was one of the main foci for the growth of national sentiment. A significant element in the unrest around 1800 was catholic, agrarian opposition to the transfer of arable land to pasture by mainly protestant landowners: 'Every fattened ox that wins his prize at St. Stephen's Green represents a peasants' levelled home' (*Irishman*, 28 Sept. 1807, quoted Bew, *et al*. 1979: 40). But the issue of land was mainly resolved before independence, the greater part of landownership being transferred into catholic hands. By then, Ireland had become an exporter of livestock, dairy produce, and excess population, and an importer of processed and manufactured goods.

However, as has already been seen, the increase in catholic control of land was partnered by massive increase in capital ownership and wealth of the protestant industrialists in the North, as certain key industries such as shipbuilding and textiles expanded with the growth of the empire. While the South, at least in the early part of the nineteenth century, continued to stagnate because of the limitations on export to Britain of land produce, the industrial North—or its entrepreneurs, bankers, and capital investors—was developing as a beneficiary of imperial expansion. Hence certain economic determinants, which partnered the religious divisions and contributed to the political polarization of Ireland, were added in the nineteenth century to more long-standing issues of a colonial nature.

The reduction of the Irish population from some eight to six million people from about 1845 to 1849 is well known. Less known is the fact that very few appear to have died from starvation, probably many more from related diseases. The majority emigrated to Britain and America. The impact the gross evil of the famine makes on our contemporary consciousness tends also to hide from it an important social structural change which occurred at the time. The famine had little negative effect on the better-off and economically dominant Ulster population and some positive effect on the growth of the middle-class catholic farmers in the South. It was in fact the destitute, particularly in the West and South-West who suffered and who were practically wiped out. It was partly as a result of the bettering ratio of land to people that the farmer-owners with 30 acres or more came to dominate the new Irish nation (Larkin 1976: 1245). As protestant landlords sold off their land to former leaseholders in order to pay their debts and mortgages, a substantial increase in catholic landownership took place. All final restrictions on catholics owning land had been removed in the relief Act of 1794, the same Act in which the small degree of suffrage had been obtained. This process of land transfer to Irish catholics, thus begun around the end of the eighteenth century, passed the crucial half-way stage by the end of the nineteenth century.

A case of relative deprivation appears to have occurred. As Larkin notes, 'Irish nationalism was rooted in something more tangible than Irish Catholics' being "conscious" of their being merely Irish. They were also aware they were economically, culturally, socially and politically deprived' (Larkin 1976: 861). The struggle for home rule and, later, for independence became equated with the struggle for political power by this growing middle class, which was subordinated both for its religion and for its colonial status. Now, it sought both the wealth of the country and the power and status which comes from self-administration. To what extent the movement became national-popular is difficult to assess. But it is clear that the land war of the latter third of the century created a bonding between the lower and middle farming classes beyond the confines of the West of Ireland where most agitation took place.

Emancipation and the greater prosperity of post-famine Ireland meant that an indigenous catholic commercial group also grew, along with a teaching profession, a locally trained and educated clergy, a small catholic legal profession, and a number of catholic civil servants. Along with the farmers, these became the real motor forces within the emerging catholic–nationalist group which came to make its eventually successful bid for power in the period from 1885 to 1921.

The Religious Change: Centralization, Conformity, and the Devotional Revolution

A cultural transformation among the catholic Irish also occurred, paralleling the impact of the Ulster protestant revival. Larkin (1976) argues that, during the first half of the nineteenth century, the personal wealth of the clergy increased. They appeared to live just above the level of the average and more prosperous members of their rural communities. The reason for the growth in their wealth does not seem clear. At the same time, there was a steady and substantial rise in the numbers of both priests and male and female religious, a fact possibly related to the rapid rise in the population at that time. The rise in the number of clergy, shown in Table 3.2,

TABLE 3.2. The Catholic Clergy, 1731–1871

Date	No. of priests (parish priests and curates)	Catholic population	No. of Catholics for each parish priest or curate
1731	1,445	2,293,680	1,587
1800	1,614	4,320,000	2,676
1840	2,183	6,540,000	2,996
1871	2,655	4,141,933	1,560

Source: Connolly 1982: 33.

was, however, not in line with the rise in population, which expanded at a much greater rate.

The church attendance of the catholic population prior to the famine was not over-impressive (Connell 1968; Larkin 1976; Miller 1975; Hynes 1978; S. J. Connolly 1982, though contested by Corish 1985; S. Connolly 1985 replicates Miller). In terms of weekly mass attendance, participation rates before the famine were lower than in contemporary Ireland. Miller (1975) interprets the census figures for 1834 to produce attendance rates in Irish-speaking areas varying from 25 per cent to 33 per cent on average, around 50 per cent in English-speaking areas, and between 70 per cent and 90 per cent in some of the larger towns. He points out that few had a long way to go to church. Despite shortages there were still enough priests to provide cover for masses. Larkin (1976) finds a much improved mass-attendance figure for the post-famine period averaging out at about 75 per cent.

Although the figures remain largely undisputed—except by Corish (1985)—their significance is not. Larkin attributes the low attendance before the famine to the prevailing folk-religion. There can be little doubt that this was strong and combined well with official catholic practices. It was also at its strongest among the lower classes and the dispossessed (S. J. Connolly 1982: 100–20, 135–74). Wakes and stations, local religious forms of 'the enchanted world' variety such as prayer attendances and penances performed at holy wells, the seeking of cures and blessings particularly on the large number of feast days which were held in popular esteem, were the stuff of daily life. With the famine, folk-religion was largely lost and replaced by the devotional revolution, outlined below. Miller applies a particular sociological explanation to this change, adopting Nottingham (1954; 1972). The transformation is seen as a passage from a pre-literate, peasant, stable, and mainly undifferentiated society with a folk-type religion, to a pre-industrial, changing, and more-differentiated society with its separate religious and social institutions. However, Miller holds that the two religious forms probably coexisted for some time until the 1840s. The trust in folk-religion broke down when the harvests failed and the famine ensued. Religious practice in formal catholic terms was then turned to more forcibly because it protected against rootlessness. One can add that the church was also coming to be seen as on the side of the oppressed. It was battling for land for the people as well as for improved religious commitment.

As Table 3.1 indicates and Hynes points out, pre-famine Ireland was hardly an undifferentiated society.[7] One aspect of Miller's interpretation thus has to be reworked. In fact, several factors probably have to be considered. It is quite possible that cottiers and the very poor did not go to mass at all, or in rota, according to whose turn it was to wear the only pair of shoes or decent set of clothes in the house. Also, because the destitute were all but wiped out by the famine, the percentage attending mass would have improved automatically. According to Lee (1973) 40 per cent of the cottier- and labouring-classes disappeared during the famine and a further 40 per cent over the next sixty years. In addition, as Hynes himself argues, the dominant class most associated with the Roman catholic church, the better-off farmers, who were developing the bourgeois attitudes of treating land as both capital and commodity (MacCurtain 1974) were precisely those whose numbers increased over the period, as land transfers to them continued. The very antagonism which existed between the farming and cottier classes could well have had a religious dimension, with lack of appropriate dress being additionally ridiculed by neatly dressed farmers and their families, reinforcing the traditional values of decency among the poor (Bowen 1978). As Connolly (1984) has described, class and status distinctions came right into the church, particularly in the allocation of the best pews to the more prosperous classes.

Finally, the decline of these practices has to be related to the positive intent of a growing number of clergy to combat evangelical proselytism with Tridentine religion and devotionalism, in addition to their commitment to fight for the disestablishment of the Church of Ireland. The Irish second reformation or evangelical revival occurred during the first half of the nineteenth century, as already mentioned. On the Dingle peninsula,[8] for example, work by an evangelical minister and his helpers resulted in the establishment, from among the former and mainly catholic poor, of a Church of Ireland community of 800 persons by 1845, including one former catholic curate. The work of a previous evangelical minister who

---

[7] Hynes, however, claims that Miller's own figures show very high mass attendances before the famine, of the order of 70%; this may be true of some of the parishes Miller mentions, but hardly squares with the overall figure Miller's statistics give.

[8] For an account of the Dingle case see J. H. Murphy 1984; one must, however, remember that the Dingle case appears to be one of the few areas in the West of Ireland where there was any substantial success at proselytism by the evangelicals, another being Oughterard, Galway, which Murphy also covers.

had had limited success before the famine now developed considerably under the famine's impetus. Such converts were known as 'soupers' by the catholic population, on the grounds that they had converted for material benefits and had been bought out of poverty. The basis for the accusations lay in the benefits of schools, welfare, and the membership of farming colonies which converts enjoyed, though such membership only provided a pittance of an income.

The reaction of the catholic leadership to proselytism took the form of a general strengthening of clerical discipline and an attempt to renew popular catholicism on the Tridentine model. One way was to develop the parish mission, which has since been a feature of Irish catholic spiritual life. The Irish Vincentian order was formed during the 1830s precisely to perform such a task. For one or two months—the practice today is one or two weeks—a group of priests would descend upon a parish and seek to renew the faith of the people through an intense programme of prayer, preaching conversion, and promoting the sacraments. In the Dingle case, the mission was eminently successful, despite the initial opposition of the clergy. Later, other aspects of Tridentine reform were introduced in the parish, as confraternities and a Christian Brothers' school were set up. But one of the consequences of the success of the mission was the widening of the gap between catholics and protestants in the area, as the missioners denounced those who had turned away from the faith. Also, the process of a spiritual renewal in part based on anti-protestantism heightened this awareness. As the Vincentians' provincial wrote in 1846,

> At last ... the object of our mission has been accomplished. The spread of heresy has been halted, attachment to the faith has been strengthened, even the most unenlightened have been instructed, a truly catholic spirit flourishes, confraternities have been begun to assist the poor and teach the young, the dangers from the apostates have been overcome, and the unity of all ranks of people for the defence of religion is now certain. (Letter by Philip Dowley to the superior general of the order in Paris, quoted Murphy 1984: 167)

The general social impact of the mission is almost gloated on by the mission's leader, Fr. McNamara, and is most telling of the sociopolitical divide:

> The magistrates and the gentry were taken no further notice of as we appeared. This was especially remarkable in the case of Lord Ventry, the

great local proprietor who was regarded as a demi-god by the people of Dingle and its vicinity. We had occasion to observe . . . how . . . everyone uncovered to him and bent lowly before him . . . But as the mission advanced their respect was quite absorbed by the missioners, and the great territorial potentate ceased to appear, being unable to brook the humiliation. (MS, quoted Murphy 1984: 167)

The reaction to protestant proselytism was thus combined with both a general renewal of post-Tridentine catholicism and a growing political religious opposition to the establishment of the Church of Ireland. From 1852, further religious sentiments worsening relationships with protestants were added to these three. They were fostered by Archbishop, later Cardinal, James Cullen (1852–78), first as archbishop of Armagh and later of Dublin. With him came other aspects of religious renewal which had been spawned outside of Ireland and imported into it by that part of the catholic clerics which had been formed on the continent. Cullen Romanized to a significant extent both the spiritual life of the clergy and the devotional life of the people. His institutional and spiritual reforms were wide-reaching. The worldly life of the clergy, already under attack before Cullen's arrival, was transformed under his guidance. But besides helping to strengthen the regular spiritual practices of mass and the sacraments already in progress, Cullen also helped to expand the devotional practices we so frequently associate with the stereotype of Irish catholicism: the cult of the saints and the Virgin, rosaries, indulgences, scapulars, images, novenas, vestments, and holy pictures, practices now largely in decline throughout Ireland.

At the same time, Cullen had to ensure the conformity of bishops to the new Roman discipline. Irish bishops were known for their independent minds and the four archbishops of the four provinces had sometimes been at loggerheads in the recent past. One such case was the dispute between them in the 1830s whether or not to participate in a national school system for Ireland funded by the government. To improve unity, Cullen succeeded in a number of institutional changes such as the selection of bishops. Much local power in this procedure was lost, as Rome became more active in their appointment. Newly appointed bishops thus tended 'to be ultramontanes, because Rome was not only the theoretical but the actual source of their own and Cullen's real power in the Irish Church' (Larkin 1976: 648). Hence, a more united, island-wide hierarchy began to come into existence, laying the basis for a dialogue between the religious and political leaders of the catholic nation in the latter party of the century. The dialogue was reinforced by the eventually successful political religious opposition of the Cullenite church to the Church of Ireland establishment. The blow for justice here, achieved with the disestablishment of the Church of Ireland in 1869, was both a blow for the Irish people and a blow for the renewed Tridentine catholicism.

The Mutual Strengthening of Religious and Political Ideologies

The religious, organizational changes of the period thus led to a strong bureaucratic, authoritarian, centralized, and devotional structure in the Irish Roman Catholic church, many aspects of which became embedded in the popular culture of emerging catholic nationalists. The whole development could only appear abhorrent to protestant revivalist counterparts, strengthening their anti-catholicism and detestation of Rome. It may well have helped the different classes of the two main reformed churches in the North of Ireland, the Church of Ireland and the presbyterian, to become more politically united, evangelically oriented, and aggressive.

By now, the middle classes, particularly the farmers, had moved to the forefront of the lay–clerical alliance. A link had already been there for some time. From at least the late eighteenth century, the farmers were those who were closest to the church, besides being the main source of candidates for the priesthood (S. J. Connolly 1982: 35–40). As the political emancipation, spiritual renewal, and prosperity of these better-off groups developed, so did the tie between the emerging national creed and institutionalized religion within their common consciousness. Irish Catholicism was a popular religion which was increasingly tied to perceptions of justice and the nation. Religious self-assertion and the achievement of catholic rights were fast becoming a form of national expression.

The effect of church teaching on the catholic family economy and culture was significant. Late marriage and the denial of extra-marital sexual intercourse were a result of both commitment to church teaching and to material aspirations. The higher standard of living sought by those in a position to obtain it required the maintenance of small families, the postponement of marriage until an adequate supply of land was found, and the avoidance at all costs of illegitimate sons who might claim the land (Hynes 1978; Connell 1968; S. J. Connolly 1982: 186–94). For the farmer to be successful, the maintenance of a stem family system[9] was required, with the inevitability of deferred marriage and lifelong celibacy for some or many family members. This ethical and economic practice also helps explain the celibate attitudes of unmarried females. To tarnish their reputation would have meant an end to the possibility of marriage. The aspirations and familial practice explains the subordination of daughters to fathers and brothers, on whom they relied for their eventual dowry (Hynes 1978: 148–9). In this way the authoritarian aspect of the traditional Irish catholic family was developed.

It is true that catholicism in Ireland did retain more liberal aspects. However, the days of an ecumenist such as Bishop Doyle of Kildare and Leighlin, who put forward proposals for the reunion of the churches in the 1820s, were numbered. But the recalcitrance of Cardinal Cullen's fellow bishops to come to heel continued until their inevitable replacement at death, sometimes late on in the century. At the level of lay political activity, some forms of liberalism survived, though perhaps not as much as is sometimes claimed. Though the parliamentary tradition was vigorously adopted by Daniel O'Connell, he was solely the leader of the Irish catholics. But this fact was also combined with a related attitude. He found great difficulty in considering the protestants part of the Irish nation. This was despite his intent, clear from his early political career at the turn of the century, to create an Ireland in which all denominations would be united in a single nation. Boyce's (1982) analysis of O'Connell here is very subtle. While recognizing the ideal for which O'Connell strove, Boyce points out the contradiction between O'Connell's liberal outlook on the need for equality between religious individuals, and his view that because the catholics were the majority, they would 'replace Protestants in wealth, property and position in Ireland' (Boyce 1982: 144). As will be seen later, the problem of defining a national culture in the terms of the majority has turned out to be a continual one in both Ireland and Britain. It has a particular ideological function in maintaining the opposing groups of catholic nationalists and protestant loyalists.

---

[9] The stem family system is the one whereby inheritance is channelled through the eldest son. Stem family households typically consist of parents, their eldest son and his wife and children, and his unmarried brothers and sisters.

O'Connell also regarded the catholic church as the genuinely national church. He opposed the setting up of what he termed the godless university colleges of Belfast, Cork, and Galway because he opposed mixed protestant–catholic education (Lyons 1973: 106). Even if he showed dimensions of liberalism in his speeches to protestants and mixed groups, he did confirm the expectations of audiences made up solely of catholics, whose concept of politics was the defeat of protestant domination and the imposition of their own (Boyce 1982: 143–6).

There are two further ideological aspects of Irish nationalism which go beyond the catholic religious element and were to be handed on to posterity. One of them, the republican tradition of violence, was to pose a threat to the development and retention of a catholic–nationalist alliance and hegemony. The other was the Gaelic revival and developed links with catholicism. The revival of language and culture which developed in the last third of the nineteenth century, happened across denominational boundaries. Anglo-Irish gentlemen and intellectuals had long appreciated it and took a prominent part in its return. But the movement did not retain this trans-denominational impulse. The catholic clerical leadership was already beginning to form an alliance with the gaelic league from its inception in 1893. There was some mutual aid to be had in obtaining the aims of both groups. Bishops found support from league leaders in attacking the central control of the education board. For their part, league leaders found it necessary to ally with the church in order to achieve the aim of teaching Irish to every child, as the manager of the local catholic Irish was the local priest. At first, the league sought to present a non-denominational image, particularly as many of its first supporters came from protestant Dublin salons, as did one of the founders of the movement, Douglas Hyde. But, already by 1899, the die was cast. Many of the membership openly opposed an attempt to associate with the Pan-Celtic General Alliance, thus strengthening the already established link of the language to that nation which was already catholic. One of the leaders of the movement, D. P. Moran, then became editor of a new weekly paper, *The Leader*, which immediately pursued a cultural and moral clean-up campaign identifiable with current catholic preoccupations: books and weekly papers from England, dance-halls and other aspects of imported 'west Britonism'. Moran summed up the catholic and nationalist dimensions of the language and cultural revival:

If an English Catholic does not like to live in a Protestant environment, let him emigrate; if a non-Catholic Nationalist Irishman does not like to live in a Catholic atmosphere let him turn Orangeman, become a disciple of Dr. Long, or otherwise give up all pretence to being an Irish Nationalist . . . We can conceive, and we can have full tolerance for a Pagan or non-Catholic Irishman, but he must recognise, and have respect for the potent facts that are bound up with Irish Nationality. (27 July 1901, quoted Miller 1973: 41–2)

Douglas Hyde himself was forced to come to terms with the power which the same religious-nationalist alliance would lend to the movement, even though he appeared reluctant to develop Moran's conclusion:

The fact is that we cannot turn our back on the Davis ideal of every person in Ireland being an Irishman, no matter what their blood and politics, for the moment we cease to profess that, we land ourselves in an intolerable position. It is equally true, though, that the Gaelic League and the Leader aim at stimulating the old peasant, Papist aboriginal population, and we care very little about the others, though I would not let this be seen as Moran has done . . . (Hyde to Lady Gregory, 7 Jan. 1901, quoted Curtis 1968: 147 and Hepburn 1980: 65)[10]

Hence, the cultural revival was to be hated by Northern protestants who saw it for what it had become: a nationalist, predominantly catholic, movement, seeking to provide a language and culture alternative to and in opposition to that of the colonial ruler, who was the protector of *their* cultural heritage, faith, and privileges. Ulster protestants of all classes wanted no part in that and its existence only served to strengthen their pro-British identity and sense of isolation from the rest of Ireland.

The other important ideological aspect of nationalism was the republican traditional belief in the use of violence to attain national goals. The myth of the phoenix rising from the ashes was to become the republican symbol of the new and armed Irish liberator, rising up from the sacrifice of blood of those who died for Ireland. The hunger strike to death was to become incorporated into this symbolism. Of significance is the seeming contradiction of those catholic republicans who on matters other than the inflicting of death on self and others showed high obedience to church discipline and moral codes. Catholics participated in the 1798 rebellion, murdered landlords who turfed out tenants, and murdered tenants

---

[10] On Hyde, see Lyons 1982: 35–46.

who replaced such ousted families. These acts were 'reserved sins', that is they could not be absolved through the ministry of a priest but only through a bishop. During the land wars of the latter part of the nineteenth century, various forms of physical violence were frequently taken up. The lower clergy themselves were often involved in the agitation, even if not generally in its more extreme forms. The Fenian movement, which grew out of the rebellious Young Irelanders of 1848, was a body supra-denominational in membership but principally led by academics and literary figures, with Thomas Davis, a protestant, as leader. It sought a totally independent Ireland, not simply an Irish parliament, and justified the use of violence to achieve its aims. It, too, was finally condemned by Rome in 1870 on the grounds of its revolutionary and secret nature. The attacks on Fenianism during the 1860s from the Irish church came mainly from Cardinal Cullen, who also saw the movement as reflecting the national revolutionary movements of Italy of the period which, as a supporter of papal temporal power, he totally abhorred.

The clerical church was also antipathetic to the use by the early Sinn Fein of civil disobedience and parliamentary abstentionism, and later condemned Sinn Fein and support for its cause in the Irish civil war (1922–3) following Southern independence. The church always sided with law and order whatever its source. This general attitude still prevails in modern catholic nation-states. But in Ireland, the church's opposition did not stop loyal catholics from disobeying on nationalist grounds, including de Valera, the leader of Sinn Fein during that war and later of the Fianna Fáil party. The continuation of this conflict between the religion of the clergy and the religion of the people will have to be partially addressed in the following chapter.

The Clerical-Political Alliance

The final solidification of the catholic–nationalist ethnic grouping about this time appears in two principal developments. Firstly, in 1891, the Irish parliamentary party ditched its protestant leader, Parnell, because of his involvement in a divorce scandal. In the fall of Parnell, church leaders played a prominent part. The leadership of the nation by one who was judged an immoral man was considered out of the question. It appears to this writer as precisely the point in Ireland's history where any hope of incorporating a protestant element into nationalist practice was finally dashed. However, and this is the second development, the parliamentary party survived the departure of this man who had constructed the alliance between the religious and political leadership of the predominantly catholic Irish. The fall of Parnell was a sequence of events of relatively short duration. But the alliance which he concluded was the result of a strategy worked out over a long period and requires our further attention.

Both bishops and priests appear to have been only too anxious to assist O'Connell in his attempt to achieve greater political power by constitutional means for the catholic population. They performed the important functions of providing meeting-places, presiding over meetings, and getting the voters out to support the cause. With the success of the emancipation movement in 1829, and despite early church opposition to his immediate steps for the next campaign, the repeal of the Union, O'Connell found the clergy still prepared to assist politically in the development of an Irish party. The church stood to gain from O'Connell's alliance with the Whigs, soon to become mainly known as the Liberal Party, through the abolition of tithes to the established church. In effect, the clergy too were the only reliable source of free advice to politically uninformed laity and were also asked frequently for their opinion even out of simple deference for their views. At the same time, the clergy were at one with the people in their fight for political and economic rights against the ascendancy, though their motivation, in all likelihood, had an additional, deeper theological level. De Tocqueville noticed of five bishops and several priests with whom he dined in Carlow in 1835

Distrust and hatred of the great landlords; love of the people, and confidence in them. Bitter memories of past oppression. An air of exaltation at present or approaching victory. A profound hatred of the Protestants and above all of their clergy. Little impartiality apparent. Clearly as much the leaders of a Party as the representatives of the Church. (1957: 130)

The local clergy were often the local organizers of the party, and bishops sometimes presided at the speeches of the leadership, arguing their way out of any apparent contradiction with Roman directives to stay out of politics. It is at this point that one can observe the political–religious conflict both within the Irish catholic clergy's mind as well as between them and Rome. The conflict within was in part caused by the refusal to accept the political nature of their own religious commitment. On the one hand, the bishops pledged themselves, and recommended their clergy in 1834, to stay out of politics and attend to their spiritual duties (Synod of Irish Bishops, Dublin, 1834; quoted MacDonagh 1983: 94). On the other, they continued their political activities of presiding at meetings and speaking out against the Tory or Conservative Party, which supported tithes to the established church. In 1839, the prefect of Propaganda Fidei then instructed the Irish primate to get the bishops out of politics. Yet in 1840, with the renewal of O'Connell's campaign for repeal, two-thirds of the bishops returned to the political arena, many of them becoming members of the Repeal association. Also the majority of the clergy were active at the local level.

The bishops excused clerical political activity mainly on the basis of practical need, or the non-pertinence of applying the general principal of non-interference in politics to the particular circumstances of Ireland. However, the particular circumstances appear to have lasted throughout the century as a whole. The conflict tends to show up the ideology of a political-free religion, which is arguably never possible. What the clergy then, as often now, failed to realize was that there are levels of politics, in some of which they can choose to be active or not, and others which are essential to their religious belief and practice.

The second contradiction appears in the conflict between the bishops and Rome and is based on the same premiss. After the prefect of the congregation of Propaganda Fidei, Cardinal Fransoni, had failed to make an impression on the political activities of the Irish bishops, the Pope himself, under pressure from Britain and Austria, forbade in a rescript such political activities. But even his attempts were to no avail. The way out for one pro-repeal bishop was that the Pope did not understand the conditions in Ireland, and therefore the rescript had no force. Perhaps nearer the mark, showing an overt political religious conscience of some degree, was the position of Bishop Cantwell of Meath. He affirmed that the basic commitment to matters of poverty and human dignity were part of the priestly calling, implying both that the Pope was beyond his jurisdiction and that religious commitment was a political one as well. Cantwell was probably something of an exception in this respect. It does seem that the other bishops justified their position by evading the central issue, and still wished to believe that their role was politically neutral. They were, then, accepting the principle which Rome was trying to assert. But it does appear from a distance that the real issue was another, namely who was to be the genuine subject of political activity, the Irish, including their bishops and priests—perhaps one should say especially their bishops and priests—or the Vatican as the prime executor of a centralized religion which sought the standardization of religious discipline.

This battle between Rome and the Irish hierarchy was renewed again in the 1880s, when the church approved and supported the Parnellite party's plan of campaign. This was a strategy for boycotting payments and services to landlords in an attempt to modify landlord rights of property to the benefit of tenants. Pope Leo XIII, strenuous defender of property particularly after the takeover of the Papal States by the new Italy, condemned the whole affair as immoral in 1888. There was never any real danger of the Irish bishops accepting the condemnation. Twenty-eight of the thirty bishops replied to the Pope, 'that the price to be paid for making his will effective in Ireland would be the loss not only of their own power and influence but of millions of Irish Catholics at home and abroad' (Larkin 1979: p. xx; 1975). They therefore did not enforce the condemnation and tried to get the Pope to make amends for his gross error of judgement. Yet neither party appears to have been able to admit the political nature of the stance of either or both. The particular grossness of the error by the Pope lay in the Vatican's failure to see the substantial political and religious change which had occurred in Ireland. To proceed on the basis of its non-occurrence was to alienate from Rome Irish catholics both clerical and lay.

The middle years of the nineteenth century had been characterized by a convention between the British state and the Irish Roman catholic church. This convention meant that the church would support the state's monopoly of physical force so long as the state supported the church's claims to legitimate interests in certain areas, particularly the right to develop its own schools and colleges with state support. It was against the background of this unwritten convention that a separate catholic primary school system was achieved under the guidance of Cardinal Cullen before his death in

1872. But by the 1880s, this convention had been supplemented and partially supplanted by a further one with the emergent catholic–nationalist grouping as represented by the Irish parliamentary party. Under this arrangement, the Irish catholic church leadership was committed to supporting the Irish parliamentary party's commitment to home rule and to recognizing that party as the legitimate and genuine representatives of the Irish people, while the party recognized the church as having a legitimate and wholly proper claim to the control of its own separate education system. So, as the Irish catholic political movement moved in the direction of home rule and national independence, it came to an arrangement with the institutional church in order to mobilize the people and legitimate the party's aims. Despite nationalist sympathies among the clergy, there was, as has been seen in Cullen's reaction to the Fenians, considerable fear of nationalism. Though already an effective strategy with Daniel O'Connell, the policy took a clearer and more decisive turn in the 1880s, as home rule came within reach. The church even obtained for its clergy a role in the selection of parliamentary candidates and, at the same time, as has been seen, officially endorsed the home rule strategy of the new leader of the party, Parnell.

The compact was mutually beneficial. Both parties could see the possibility of their principal aims being accomplished. Of course, one must remember that the church was the only organized aspect of the catholic underclass in Ireland throughout the period of subjection and into the nineteenth century, and its only real intellectual force. It is hardly surprising that it developed as the very source of national identity for its members, persecuted at once both politically, economically, and religiously. Thus, it really was for the clerical church to legitimate in a real sense any emergent political force from within the ranks of the faithful. Considering the problems and the ideological constraints of the time, the leadership did not do too bad a job. But it did operate within the context of hostility between protestant and catholic churches and was led by its own Tridentine religious policy to the unintended consequence of supporting and fostering the growth of a national identity which was to be ultimately divisive for Ireland.

Hence what Gramsci recognized as the key strategy of the church in Italy to achieve in the modern world a new form of hegemony in catholic states, came to be also the key strategy in Ireland, though within its specific historical conditions. The strategy was monopolistic in character for those areas of society which it considered its own proper realm. The issue of the rightful place of a protestant tradition in Ireland was simply invisible to catholic perception.

The dramatic increase in the power of the new catholic–nationalist party which emerged with the extension of the franchise in 1884 was due only in part to the huge increase in its parliamentary seats. The strengthening of the catholic–nationalist bond also came because the church leadership thought home rule might really be delivered democratically, without violence, by the alliance between the Irish parliamentary party and the English liberals led by Gladstone. Gladstone was now both committed to home rule and in need of the Irish party's support if his party were to retain power in Britain.

The process of legitimating changes in the Irish parliamentary party's political goals and strategy by the church was normally a *post hoc* strategy but was none the less essential for the maintenance of national solidarity. Also, when partition and national independence came, the definition of the proper sphere of the church within the new state appears almost automatically to have been enlarged. As will be seen, it began to appear that the church was going to look after all those things which concerned the moral and spiritual welfare and development of the Irish people—that is, the catholic Irish and anyone else who happened to be within the boundaries of the Irish state. The state would concern itself with all other matters, whatever they might be. In a sense, the terms of what made up either heading were very much to be decided by the church, save perhaps the basic structure of democratic government. This was built into the arrangement by O'Connell, but the contradiction to agenda-setting by the church which it implied was not yet apparent. It could be argued that, in some respects, it is still not apparent to a number of the participants in current church–state politics in Ireland.

THE BREAK-UP OF IRELAND, 1905–1922

With the formation of the unionist alliance in 1906, a settlement incorporating Northern protestants into an Irish home-rule structure had become impossible. This irreversible polarization appears inevitable in hindsight, but most Irish nationalists would not have been able to see events in that light. Loyalists had already begun to drill in preparation for civil war. In 1911, the religious solemnization of the protestant–loyalist alliance took place. The British liberals, in need of Irish support to maintain their power, yet again proceeded towards a home rule Bill for Ireland. This one seemed destined to succeed. The new leadership of the protestant loyalists, Edward Carson and James Craig, called for a unionist and Orange demonstration, which took place at Craigavon. Fifty thousand unionist and Orangemen, 'humbly relying on the God whom our father in days of stress and trial confidently trusted', committed themselves to a 'Solemn League and Covenant'. This bound those signing to refuse recognition to a home-rule parliament in Ireland should one come about. They committed themselves to setting up a provisional government in the interim period between involuntary separation from Britain and their welcoming back into the union. It is important to note that only covenant politics and beliefs could extricate Northern protestants from the dilemma of a united Ireland or betrayal of Britain. Unionist hopes were raised further when the British establishment in the shape of the Tory Party committed itself to the unionist cause. The Ulster Volunteer Force was formed and began to arm to defend the counties of Ulster from home rule. In early 1914, the officers of the British–Irish army stationed at the Curragh, thirty miles from Dublin, gave notice that they would not obey orders to quell a unionist rebellion.

As 1914 wore on, a war between catholic nationalists and protestant loyalists seemed inevitable as home rule became imminent. But when it did in September, the liberals had already modified the Act to enable individual Ulster counties to exclude themselves from home rule for a period of six years. This may have been done to avoid a cataclysm within Britain itself, but it hardly solved the problem for those living in Ireland. However, home rule did not come into force because of the invasion of Belgium by the German army and subsequent declaration of war on Germany by Westminster. The war was to turn the mutinous and rebellious protestant loyalists into virtuous and brave defenders of the empire, and Irish catholics into rebels instead.

The Irish Volunteers had been formed in 1913 to counter by force of arms an Ulster secession and to further the cause of Irish independence by violence if necessary. Shortly before their formation, the Dublin workers under James Larkin had also formed a small citizen army. When the leader of the Irish parliamentary party, John Redmond, took the Irish Volunteers into the war in Europe, the rank and file split, with the minority objectors, together with Larkin's citizen army, becoming the nucleus of forces for the Dublin rising of Easter 1916. Hardly popular with catholic nationalists as a whole, the rising was supposed to be island-wide but was badly organized. The only action was in central Dublin on Easter Monday, with the establishment of control in the city centre, and the declaration of an independent all-Ireland state. The rebellion was crushed within a week, and many leaders shot dead throughout the month of May.

It is generally held that it was this prolonged counteraction by the British government and army against the rebellion's leaders which solidified the separatist nationalism of the catholic Irish. But then the British and Ulster loyalists had felt that they, too, had been stabbed in the back while fighting the foreign enemy in Europe. Group polarization was complete and the fact that about 100,000 Irish catholics fought and died on the battlefields of France and the coast of Turkey became less and less seen as a sacrifice for catholic Belgium and increasingly as a wasted sacrifice for an oppressive empire in a capitalists' war. With the end of the war in November 1918 and the calling of a general election, the old parliamentary party of Redmond was heavily defeated and Sinn Fein, responsible for the rising of 1916, swept the board. Instead of going to Westminster to take their seats, the successful candidates set up a provisional government in Dublin, meeting for the first time in early 1919. But they had no hope of ever gaining control of the whole of Ulster, whatever the means they might be prepared to adopt. At the same time, the leadership of the protestant loyalists united around the realization that, if they accepted a six-county Ulster for their domain, they would be able to defend it, especially as the link with Britain now appeared rock solid: their sons had, after all, died for the empire. Northern Ireland, a protestant state for a protestant people, came finally into existence by the Government of Ireland Act of 1920. The other provision of the Act, a separate parliament for the rest of Ireland, was soon made a dead letter by Sinn Fein who made government and the British presence in the South impossible. However, Sinn Fein finally split on the republican question, with the majority agreeing to independence within the empire and to the crown as head of state. This was partly agreed and partly imposed by treaty with the British government the following year, 1921, and the catholic–nationalist state of 26 of the 32 counties, the Irish Free State, was created.

4
States, Religions, and Ruling Ideas

Catholic Nationalists and Protestant Loyalists Today

STATES AND RELIGIOUS INSTITUTIONS

The chief purpose of the chapter is to outline how the two opposing groups in Ireland developed their dominant beliefs in the context of the new states of Ireland, and moulded the states to support them, strengthening antagonisms between the two groups. What changed with the partition of Ireland in 1920 was that the two dominant alliances were internally reorganized. In particular the protestant bloc lost the Southern protestant landowning gentry. The two alliances were ascribed territories in which to produce their social structures, political institutions, coercive bodies, informal and interpersonal networks of maintaining influence, and more clearly defined hegemonic cultures. The two societies gave expression to their own dominant interests as far as possible unhindered by each other's interference. The process was assisted in the North by the self-exclusion of Britain from internal Ulster affairs by setting up the Stormont parliament which survived from 1922 to 1972. Protestant–loyalist domination, successor to the ascendancy, was circumscribed to one corner of Ireland while the catholic–nationalist alliance asserted total domination in the remainder.

The Catholic–Nationalist State

The direct consequences of this dramatic political change differed in each state, but paralleled each other, consolidating each alliance's power. The Southern state became more catholic and nationalist in character. Its protestant groups declined in numbers, wealth, power, and cultural importance (K. Bowen 1983; Walsh 1970). The Northern state became wholly dominated by protestant loyalists and became extremely coercive in its dealings with the substantial catholic–nationalist remnant within its borders. It was the failure to incorporate the catholic–nationalist remnant into the consensus of the Northern statelet which resulted in sporadic violence, institutionalized and structural on the one side and insurgent-guerilla on the other. The two alliances succeeded in creating sufficient structures of hegemony and coercion to maintain social control for a significant number of years. But Northern reliance on force, necessary if the character of the state was to be absolutely preserved, was the weakest point and the eventual undoing of the apparent stability achieved by the partition settlement.

The Southern protestants have diminished greatly in numbers since 1911, because of the loss of life in the 1914–18 war, some emigration, and probably mixed catholic–protestant marriages (Walsh 1970; White 1975; Ch. 7, below). In 1911, the protestant population in the South was 10 per cent. By 1925 it was down to 7 per cent and in 1971 it stood at around 4.1 per cent. Over the same period, the Church of Ireland experienced continuing rounds of church closures and became top-heavy in numbers of clergy. Statistics up to 1971 showed it as having an old-age structure and to have been in continuing decline, though there are recent signs of improvement (Census of Ireland 1981). Southern protestants do not appear to be particularly oppressed by the state (White 1975: 169), but they would clearly change aspects of the constitution and legislation if they were in a position to do so. They have always retained a low political profile. They have also never been perceived by the state as a threat, perhaps because of the revolutionary pedigree of some of their élite. As such, their own life-styles, societies, and schools have been relatively protected, but perhaps at the expense of political action.

A practical ideology for a catholic Ireland was worked out and the aim of a society uncontaminated by the outside world was pursued. Both church and state functioned symbiotically. For his part, de Valera, prime minister for most of the period from 1932 to 1959, pursued isolationist policies, economically, politically, and culturally, seeking at once self-sufficiency, neutrality, the restoration in part of Gaelic culture through the minimal enforcement of the Irish language, and the control of media consumption.

In concrete terms, de Valera was acting on behalf of the established economic interests of rural Ireland, in particular the small landowners and small businessmen. The commitment to the eradication of rural and urban poverty, while not entirely forgotten, was subordinated to the goal of economic independence from Britain. The path pursued particularly matched the clergy's vision of the profanity of towns and sacrality of rural life. This was despite the fact that the reform of Irish catholicism in the nineteenth century sprang mainly from urban areas. One must also remember that the bulk of the clergy both in Ireland and in the remainder of catholic Europe came from rural backgrounds, where the sense of God is, or was, part of the air one breathes (Acquaviva 1979).

From de Valera's point of view—and the majority of the nation and its politicians would have concurred—the fact that the Free State was over 90 per cent catholic meant that its moral and social outlook would reflect catholic beliefs. This was seen to be the democratic process at work.

The results of isolationism in terms of human suffering were massive emigration (some 20,000 every year to Britain up to the 1960s), and industrial and commercial underdevelopment until the late 1950s. But there were undoubted benefits to both church and state in this set of policies, for they were to the mutual advantage of nationalism and catholicism as these were perceived by their respective institutional representatives, namely the politicians and the hierarchy. They were the result of catholic–nationalist success in establishing a separate state, so separate that the protestant tradition as a political force had been excised from the thinking of the entire catholic grouping in the South, from church to state to popular consciousness. This occurred despite the continuity of the catholic hierarchy in Northern Ireland and despite the war experience of those catholics who had been involved in the army. Armistice Day is still only celebrated in Southern Ireland by protestants: as late as the early 1970s, you could still tell a protestant by his poppy on 11 November. The commemoration of the dead in the First World War was seen as a triumphalist commemoration of those who died for Britain. The very existence of the catholic state in all but name, the realization, if only in part, of the territory of Ireland as both catholic and nationalist, had a significant and continuing impact on both the clergy's practical theology and the laity's day-to-day perceptions of social reality. Similarly, the *de facto* existence of the protestant enclave in the North and its pursuit of protestant dominance in that state may have even reinforced a similar, though integralist, approach in the South.

With independence the Irish Free State developed a type of political party organization on modern lines, but without the genuinely class-based politics one finds in other European countries. The party structure followed was based on the Irish civil war antagonism of 1922–3, with the pro-Treaty party more or less in power until the anti-Treaty party entered politics in 1930. However, both sides developed a centralized party organization, with politicians following the party line.

The Protestant–Loyalist State

The condition of the Northern catholic–nationalist minority has been very different from that of the majority in the South. The Northern minority has been feared throughout the period by protestant loyalists for two main reasons. On the one hand, force has been required to subordinate it, and, on the other, it has always threatened to outbreed protestant loyalists, an outcome which has only been avoided by catholic–nationalist migration over the past seventy years. The loss of numerical superiority by protestants would result in the collapse of their statelet. Disappearance of their dominance would lead to a unified Ireland or else the restructuring of a new statelet with smaller geographical boundaries in which their dominance was again reassured. In any case, repartition is not out of the question. Recent work on the religious demography of Northern Ireland by Compton and Power (Compton 1985; Compton and Power 1986; Ford 1986; Commins in Clancy, *et al.* 1986) indicates a separating out of protestant and catholic, with the catholic population drifting westwards and vice versa.

The majority of the Roman catholic remnant in the North still believe in a united Ireland. They demonstrate this partially by their support for the two nationalist parties, Sinn Fein and the Social Democratic and Labour Party, though support for the latter does not necessarily imply nationalism. Despite the official socialist tendencies of these two parties, a substantial number of their voters do not appear to question the maintenance of a capitalist-type class structure in a new Ireland. The Workers' Party[1] still fails to command many votes in elections north or south of the border. The provisional movement is divided on the issue of capitalism but, for

---

[1] The Workers' Party is the former Official Sinn Fein, which supports the establishment of a socialist all-Ireland federal republic by peaceful means.

the time at least, has an openly declared policy of achieving an island-wide socialist republic by the combined means of violence and the democratic process. The SDLP, still the main party of catholic nationalists in Ulster, has dropped much of its socialist platform and maintains a delicate balance between nationalism and pragmatism in its politics. Despite frequently heard voices to the contrary, total catholic–nationalist domination would be the likely outcome of a united Ireland unless a confederal structure were implemented. This is partly because of the structure of catholic nationalism in the Southern state and partly because the national unity sought by significant numbers of Northern catholic nationalists is still on the basis of the reversal of protestant–loyalist hegemony rather than a power compromise.

Solidarity with Southern nationalists is still a major Northern nationalist perception, despite the fact that they view their Southern compatriots at best with indifference and at worst as traitors who sold them out to the loyalists. McGreil (1977: 363–408) has shown that though Dubliners find the English more acceptable than the Northern Irish, Dubliners still seek a solution to the Northern problem within an all-Ireland state. Also, there can be no doubt that the South would intervene in the North on behalf of the catholic–nationalist minority if Britain withdrew and civil war broke out. Equally, Southern catholic nationalists prefer a unified Irish state, though a majority of 54 per cent would accept a federal one based on the present two state units. One can therefore conclude that there is a sense of unity throughout the catholic–nationalist population as a whole, though one in which certain groups, namely farming interests and the church, at both clerical and popular levels, have strategically dominated, at least until recently.

By 1914, the protestant–loyalist group was well established and ready for war against catholic nationalists should it have proven necessary. Issues of class conflict within the group were to remain largely subordinate, as the matter of the nature and defence of loyalism itself was to dominate the scene right down to the present day. There was to be an expression of a certain solidarity between members of both the catholic and protestant working classes in the Belfast demonstration of 1932 against the inadequate poor relief during the period of particularly high unemployment. But such co-operation was rare and quickly taken over by sectarian issues.

Protestant–loyalist and catholic–nationalist politics in the Northern statelet differed from those in the South. McAllister (1980) describes that political culture as traditionalist, and dominated by clientelism, parochialism, local attachment, and ascription: politicians were elected on the basis of local reputation and solely with reference to local issues. During the period of the Stormont parliament, half the seats were uncontested. This was frequently true of Westminster elections too. It would seem that the special state character of Northern Ireland, with its very legitimacy always under threat, has been the main reason for the maintenance of this pattern of politics.

Up to the late 1960s, catholic nationalists were split between two main political groupings. There was the Nationalist Party, a weak organization for which local priests had to provide some kind of legitimation. As a party, it really only exercised a modicum of power in relation to the Stormont administration. Then there were the republican parties who focused their attention on Westminster elections. The disorganized nature of catholic–nationalist politics was only turned round with the emergence of the civil rights movement of 1968 and the subsequent forming of the SDLP in 1970. However, during the 1970s, the socialist element was weakened with the departure of labour leaders. The remaining leadership represented a modified nationalism in which new middle-class moderates held the upper hand. The party as a whole tended to represent the employed and the over-thirties. The unemployed and the under-thirties have lately tended to support Sinn Fein since its adoption of an electoral strategy in the 1980s (Bew and Patterson 1985; O'Dowd, Rolston, and Tomlinson 1980; Moxon-Browne and Irvine 1989).

The Unionist Party became the central political organization of the protestant–loyalist alliance. True, there was some fragmentation of members on class issues. The occasional independent unionist and independent Orange-lodge representative in politics were a feature of politics from 1880 to 1972. The Northern Ireland Labour Party also was the principal organization representing protestant trade unionists. But none of these groupings ever came near to threatening the dominance of Unionist Party organization, much less the solidarity of protestant loyalists, which always appeared total on the issue of incorporation into a united, independent Ireland. It was only with 'the Troubles', from 1968, that unionists broke ranks. Yet, despite the splintering, they have succeeded in allying the present two main parties against the Anglo-Irish agreement of 1985.

The Alliance Party of Northern Ireland, formed at the time of the abolition of the Stormont parliament, represents those catholics, protestants, and others who disavow the twin alliance of Ireland but retain at least a practical attitude on sovereignty. They prefer British to Irish rule and seem unlikely to move from that position. While the party contains many who actively seek peace and reconciliation, it would be wrong to think of them in any sense as overcoming the basic conflictual components of bloc power in Ireland. 'Castle catholics', those who found a place in the Stormont administration and developed a stake in the union, have also voted alliance, but are probably now, in 1989, switching their allegiance to the newly found conservative associations, which are seeking to bring British party loyalties to Ulster.

The Northern Ireland statelet was founded on an alliance between the protestant loyalists of Ulster and British imperial and capital interests, particularly as represented by the conservative party. But that is what the arrangement was: an alliance rather than a consolidated interest or a genuinely shared culture. Protestant loyalists dominated the statelet from its formation. The British government only obtained certain concessions from the Stormont government in implementing measures of the welfare state variety after 1944. In a detailed study of the Stormont archives, Bew, Gibbon, and Patterson (1979) have shown that there were indeed different currents within the Stormont administration but the ones which predominated belonged to those among the ruling protestant classes who were bent on preservation of the status quo for their own purposes, including their own dominance of the protestant alliance as well as their particular sectional interests.

The range of tactics used to keep catholic nationalists subordinate was quite comprehensive. At the political level, proportional representation was abolished for local elections in 1922 and for Stormont elections in 1929. This minimizing of catholic–nationalist representation was aided by the gerrymandering of electoral boundaries for local government elections, which concentrated the catholic–nationalist vote in a smaller number of constituencies. At the level of law and order, total control of the subject population was achieved in two ways. Firstly, the Ulster Special Constabulary, invented to supplement the main police force, was derived largely from the former Ulster Volunteer Force. Divided up into three groupings, the A, B, and C special constabularies, according to diminishing power, responsibility, and time commitment, the force was soon cut down to class B only and these came to be known as the 'B Specials', a thoroughly armed, militant, semi-private, and sectarian army. It could be called upon at any time and indeed was from time to time. Secondly, through the enactment of a special powers Act in 1922 Stormont gave itself and its police force the powers of detention without trial. These were used copiously over the years of Stormont rule. Catholic nationalists could be and were arrested with little suspicion, detained, frequently beaten, and all with little or no possibility of redress. The British government over the water could have been a million miles away (for detailed accounts see Arthur 1984; Buckland 1979; Farrell 1976; for a summary see McAllister 1983; for a summary of all forms of discrimination adopted, see Whyte 1983; Hillyard 1983).

But the British government was still involved in the reproduction of antagonisms in the Ulster statelet, rather than passively accepting the perpetuation of inequalities and discrimination against the catholic–nationalist minority. O'Dowd, Rolston, and Tomlinson (1980) point out that when the centralized state comes in simply to administer and maintain law and order, it tends to reproduce relationships which pre-exist its intervention. When the British state began its policies of social interventionism from 1945, it succeeded in fragmenting the local power base of unionism by centralizing the sources of welfare and making them at least in part available across the sectarian divide. There was thus some impact on discrimination. However, when Stormont was abolished in 1972 and central government took over entirely the responsibility for local administration, it adopted the technocratic, non-policy making approach to its new role. Part of the reasoning behind the move was the search for an effective policy to contain the violence within Ireland and prevent it from spreading to Britain. It was also hoped that bureaucratic impartiality would reduce discrimination further. But the technocratic approach did not lessen discrimination but reproduced it in a subtler form. The continuation of inequality is still clear in the fields of income, housing, wealth, and employment but most of all in the protestant–loyalist alliance's embargo on the minority ever holding political office.

At least up to 1985, the British government submitted to the threat of violence from the loyalist community and accepted the protestant–loyalist veto on any change in the constitutional structure of their state. With the Anglo-Irish Agreement at Hillsborough in 1985, it has sought to lay aside that veto, and establish in principle its political right to modify the structure of power and control in Ulster by involving the Republic of Ireland in a common policy on security. However, in the interests of Western European stability and its own financial position, the British government is perhaps still putting off the final offer—either British withdrawal or some form of power-sharing with catholic nationalists—in the hope of loyalists coming round to what is seen as a more reasonable position.

The class structure of Northern Ireland is predictably dominated by protestants. Of course, with the development of international monopoly capital and multinational companies, additional sources of power have been brought into play. But it is likely that the professionals who have been responsible for the running of such companies at local level will have been allied to the more liberal group among the upper protestant classes, as represented by such families as the O'Neills, who have looked to the English public schools for the right sort of education. Wallis, Bruce, and Taylor (1986) list this mainly landed gentry among the cosmopolitans of Ulster society. Also cosmopolitan in outlook are a variety of traditional professions, particularly university lecturers. Many of these moved their allegiance to Ulster's alliance party on its foundation in the mid-1970s. The limitations on the power of these liberal groups within protestant loyalism are demonstrated by the fact that they have only been allowed to function among the leadership of the people so long as they obeyed the basic tenets and values common to the alliance as a whole. This is shown in the protestant–loyalist rejection of Terence O'Neill and O'Neillism, and of his successors, Major Chichester-Clark and Brian Faulkner, as prime ministers of Northern Ireland (1968–74).

As Eversley (1989) has recently shown, while only 2 per cent of Roman catholics hold managerial and professional positions compared with 4 per cent of protestants, they are over-represented in the bottom third of the Northern class structure, but generally within the framework of a traditional dual protestant–catholic labour market. Those members of the catholic population who sought to advance up the class structure had to adjust to more marginal roles among the professional classes. Those becoming solicitors, barristers, university lecturers, and doctors had to accept limitations on their careers if they stayed in Ulster. Catholic civil servants usually had to abandon any practical political project if they wished to proceed through the ranks. Even so, those who had any career success were extremely few in number. In terms of the acceptance of law and order, the bulk of the catholic–nationalist remnant form a part of the civil society of Northern Ireland, though as much as a third of the remnant can defect from this consensus, as when supporting the Provisionals over the 'H'-Block prison issue. But in terms of the protestant–loyalist philosophy, the overwhelming majority of catholics oppose the state's legitimacy.

This does not mean that there is no cultural convergence at all between the two alliances in Ireland. Apart from basic agreements on the necessity for law and order for survival purposes, an issue which will be treated later in the chapter, there are a number of similarities which should be mentioned. Both sides are based on capitalist activity and both governments have followed modernization rather than self-development strategies since the 1960s.[2] More particularly, despite some continuing differentiation in industrial specialisms, such as the declining presence of heavy engineering in Ulster, the types of industrial activity are increasingly similar as are the economic and unemployment problems, population growth, and productivity, with the probability of higher deficits in the North (FitzGerald 1972: 63–85; New Ireland Forum 1983–4: ii. 8–13; O'Dowd and Wickham in Clancy, et al. 1986). Also, investment patterns in both parts of Ireland are increasingly similar, being a mixture of British, international, local, and cross-border companies. Internal commerce, from heavy machinery to domestic goods, largely ignores the border, and the high street banks are the same throughout Ireland. The main economic differences reflect the different structures of the state and the different state policies adopted by the British and Irish governments at any one time, though with a significantly higher state sector of employment in Ulster.

There is likewise considerable overlap in cultural areas. There are strong similarities in inter-family and inter-gender relations. There

---

[2] For the classical analysis of modernization theory and strategy, see Frank 1971.

is a recognizably similar personal style to all face-to-face interaction throughout Ireland (for the North, see Harris 1972; Easthope 1976). There are similar strengths to kinship networks. Catholic and protestant church organization straddles the border. There are certain common features in ethical attitudes: puritanical attitudes in sexual matters, conservatism, friendliness, 'down-to-earthness', sense of duty to neighbours, critical attitudes to officialdom, and similar judgements as to what constitutes good and bad conduct (Harris 1972; Leyton 1966; 1975; Buckley 1984). Public events sometimes confirm these viewpoints, as has appeared in the continuing joint catholic–protestant political opposition to the abolition of the criminalization of homosexuality. However, the existence of a certain level of common values is no indication by itself of a ground for peace. It may even be possible to obtain approval from opposing groups for a set of moral maxims such as peace, justice, and freedom—see the lists of values in McKernan's (1980) research, for which he found a high level of acceptance among both Northern Ireland communities. But when it comes down to concrete moral judgements and commitments, such abstractions will be found interpreted in opposing ways.

Religion

The Roman catholic church retained a unified hierarchical organization for the whole of the island with its bishops meeting periodically at Maynooth. In 1927, 1929, 1956, and 1960, there were plenary synods and councils. Since then, the hierarchy has met more or less three times per year, but usually in secret sessions.

In the North, the bishops pursued the Irish catholic community's interests in what could only be called a spirit of 'pillarization'. This is the arrangement in Holland whereby various institutions such as media, schools, cultural organizations, welfare services, and hospitals are duplicated, and run by the separate catholic and protestant communities. In Holland there is even a third or state sector. In Ulster a full panoply of institutions emerged from the dual pressure of the church's concern for its people, and the Northern state's discrimination against catholics. The clerical and lay leadership also experienced a third pressure, that of protecting the faithful from the proselytizing activities of a number of evangelical protestants mainly among the young. The church obtained a separate school system, funded mainly by the Northern

Ireland government, similar to the system of catholic schools in Britain. It also set up its own scouting organizations, clubs both sporting and social, with Irish music, Irish dancing, and Irish or Gaelic games. These games were and still are much loathed by the majority within the Northern protestant community. Priests continued to act as sponsors of the local catholic–nationalist political structure, chairing meetings of the party and permitting the political use of the parish hall. The parish structure and school became the heart of the Northern catholic–nationalist community, rendering culture, politics, and religious identity as interchangeable for members of the in-group. Employment patterns, which followed the routes of enforced segregation in favour of protestants—for they controlled the greater part of the job market and especially the major industrial enterprises and the entire government sector of civil servants and police—were even reinforced by the system of job references provided by the parish priests and catholic notables. The church certainly had to follow this path in the face of the external constraints; but the practice did strengthen its status, power, and the structure of political divisions.

In the South, the church hierarchy pursued the path of full cultural control. Conditioned by church canon law, participation by clerics in politics was forbidden. But at the same time, an enormous sense of responsibility to protect the church, the family, and the minds and hearts of the faithful from incursions by the state was perceived to be the basic duty of the church leaders. The leadership of the state knew in large measure what was wanted: it was part of the national–popular consciousness as well as being itself a result of the convention between church and Irish parliamentary party in the previous century. What had been convention became part of the natural, common-sense universe of catholic–nationalist ideology. No real consultation would have been necessary on that score. As will be seen in the next chapter, when the republican wing under de Valera took over as the Fianna Fáil party in the 1930s, constitutional law was restructured, according to both a reformed republican ideology and current Roman social teaching, and in those areas where the high clergy thought it necessary. In addition the church promoted strong centralized relationships between itself and government and actively discouraged the spread of power to the periphery or local level.

Within the ranks of the clergy and religious orders themselves, there was a far from complete, centralized structure, but rather a dual one. Apart from small orders, usually of nuns, set up by local bishops for local needs, religious orders have always had considerable independence from the bishops. Part of this independence must be attributed to their ownership of houses and schools, paid for directly by their benefactors, payers of school fees, or investment funds. This has sometimes led to disputes between religious and secular clergy, between orders and bishops. For example, in the Northern context, the previous bishop of Down and Connor, Dr Philbin, refused for most of his period of leadership in Belfast to have Jesuits visiting or residing in his diocese. Not that the prohibition always succeeded. It began to be violated in the late 1960s by Fr. Michael Hurley who on one occasion addressed the presbyterian general assembly in Belfast against both the bishop's private wishes and Ian Paisley's open and defiant demonstrations.

The dual division of ecclesiastical power has also been important because the religious orders have often been the vehicle for innovative theologies differing from the Roman line. Because such religious orders are themselves controlled from Rome, Roman discipline is conveyed via a route other than by the bishops. Even so, some orders still remain slightly independent from Rome, particularly the religious orders of solemn vows, whose existence preceded the reformation. Their numbers include the Augustinians and Dominicans, both of whom have been influential in Ireland and have produced theologians of liberal or left persuasion. Despite their dependency on Rome, the Jesuits too have been important innovators on the contemporary scene in Ireland, particularly in the areas of education for ecumenism and industrial relations. Nevertheless the predominant catholic power lies with the hierarchy and secular clergy. Thus, while it is possible to identify liberal catholicism in Ireland from the late nineteenth century to the present, and though there has grown up a climate of liberal dissent and criticism of conservative catholicism, it is still possible to assert the prominence of conservative catholicism, particularly in its alliance with nationalism. The predominant political religious form was and remains, with some modifications and with increasing opposition, a centralizing and conservative catholicism.

The protestant churches in Ireland also retained their island-wide organization after the political partition of the island. The church of Ireland still has one all-island synod, the presbyterians a single general assembly, the methodists one conference. Even Ian Paisley's free presbyterians, who have communities in County Cavan and Dublin, are centrally organized, though this may reflect the practical situation in Ireland and the role of the 'big man' himself within the church he founded. The annual meetings of the churches' central bodies are a forum for religious political debate and have been known to express tensions between the values of the majority Northern and minority Southern protestants.

In the Northern statelet, the power of the main protestant churches at present remains subordinate to that of the fundamentalists and the religious political societies. The societies of the Apprentice Boys, the Black Preceptory, and the Orange order, in conjunction with protestant paramilitary groups, could still form the basis for protestant political power if Britain withdrew from Ulster. From its domination by the old ascendancy, the Orange order in particular came to embrace protestants from all classes and churches. In fact, in rural areas today, the Orange hall has frequently been the only place where protestants of all shades have met to renew their belief and commitment to Ulster protestant loyalism (Harris 1972: 162–5). It is important to note that these organizations are internally stratified. Those nearer the top, particularly in the Black Preceptory, the inner sanctum of the society structure, are there because of their moral and religious standing in the local community: wealthy farmers, businessmen, good churchmen, or at least of known moral probity. They tend to represent the 'religious' as opposed to the 'rough' end of the protestant spiritual spectrum.

Fundamentalist religious groups have increased their power in Ulster in recent decades, especially the free presbyterians. Their small church membership of around 10,000 hides the wider appeal Paisleyite politics has for the small businessmen and farmers of protestant Ulster. The setting up of a new form of Ulster Club organization involving paramilitary activity in the wake of the Anglo-Irish accord of 1985, is an added dimension to the struggle for power within the alliance. It is likely that the inner core of local church laity, such as select vestry members of the Church of Ireland and presbyterian elders, provided and still provide one of the links between the material and spiritual interests of the groups in the alliance, as these laity were and are active in the business and commercial fields also.

THE DOMINANT BELIEFS OF THE TWO ALLIANCES

As we examine the dominant beliefs of the two alliances, it must be stressed that we are not looking at beliefs possessed by each and every person who identifies to a greater or lesser extent with either set of traditions. Rather we are looking at those beliefs, supported by significant power bases within each bloc, which have won out in the political process in the past seventy years and still appear to be doing so, in some cases with much decreased vitality.

The dominant beliefs of the catholic–nationalist bloc still are that the group forms a people who are Gaelic-Irish, constitute a nation, are republican, and populate an island which has a natural, inner political unity. We reiterate: the dominant beliefs of the alliances which we have termed historical blocs are not necessarily shared by all but are promoted successfully by dominant groups and have some grounding in the consciousness of the wider membership. Thus, Irish nationalism is conceived by most members in an abstract way, but it has concrete import for key groups. Equally, a particular type of catholic identity may not belong to all but remains politically dominant for similar reasons. The outcome of the dominance of both is that historically there have been critical mediations between religion and politics for catholic nationalists.[3] As will be seen in the remaining chapters, a particular form of Irish catholicism known as monopoly catholicism has assisted in the concrete success of the mix of the beliefs with economic beliefs in capitalism, ownership of land, and inheritance, helping to bring the elements to an explicit formulation in the 1937 Irish constitution, which has ever since shown considerable resistance to reformulation.

The prevailing protestant–loyalist beliefs stress the existence of the people of Northern Ireland, who are distinguished by their Ulster protestantism and democratic values, and by their claim to a flexible territory of Ulster. Despite some socialist groups within the alliance, the dominant economic belief is capitalist. A further central belief has recently begun to appear as equivocal, namely

---

[3] In general terms, it is true to say that there is no set pattern to nationalist beliefs throughout the world. Also, at popular level, they may contain many contradictions. Beliefs which do not cohere theoretically are still relevant from situation to situation. An example from Britain might be useful here. It is possible to be British and Welsh, but Welshness might predominate in North Wales in one particular situation such as the language question and Britishness subordinated; likewise, in South Wales, particularly among the industrial working class, Britishness might subordinate Welshness over the same question.

loyalism itself. Related to this is the final principal belief element, the conviction of the legitimacy of force in maintaining the subordination of catholic nationalists within the statelet. A fuller account of these two groups of dominant beliefs benefits from some comparison between the two.

'We, the People...'

The only way catholic nationalists can fulfil their present aspirations as a people is by extending the Republic of Ireland to the entire island. This would mean containing within their state one million persons to whom such a national identity would be totally antithetical. The Irish people are defined primarily by their allegiance to the present restricted or future enlarged republic, but it is popularly understood that they will normally have the further characteristics of Gaelic Irishness. The belief is still popular that such a race exists, rather than there merely being a cultural group based on nationalist, republican, and catholic religious sentiment.[4] It is a belief which has intellectual credence in Ireland (O'Tuathaig 1986 and M. M. Ireland's responding letter, *Irish Times*, 8 Dec. 1986).

Theoretically, one would expect the Northern catholic community and the six counties of Ulster to be considered a lost remnant, unredeemed from British rule, and indeed this is by far the most popular interpretation. But, of course, the problems that arise out of this interpretation are twofold. Firstly, there is the problem of the protestant population itself in the North. Their position as Irish people is ambiguous. For the hard-line republican position, there is even a duality: protestant loyalists are taken as foreigners with an alien culture, and as either a pariah group—the mere instrument of British imperialism—or simply not there. Their culture has no rights, no claim to an alternative identity. Intellectu-

---

[4] The idea of race is based on the belief that human persons exhibiting certain differences such as skin colour or peculiarities of pronunciation are in fact significantly different at the physical level and, as a consequence, psychologically and culturally different. Thus such things as IQ scores, temperament, and a range of abilities and skills are seen as determined biologically. Sociologists and contemporary anthropologists have largely undermined such beliefs. There are no firm scientific indications for the existence of different races. Variations in skin colour are as indicative of biological determination as the difference between blue and brown eyes. Character differences are either randomly distributed throughout the population or are cultural in origin and the product of individual or group socialization processes. For summaries of the debate, see Banton 1987; Miles 1989.

ally, the problem is the British presence in Ireland. But emotionally, the non-people are hated or at least detested. Clare O'Halloran (1986) has shown how the social reality of Northern Ireland remained a continuing blind spot for Southern Irish nationalists particularly from the founding of the state to 1949. We aim to show in Chapter 5 that this blindness was in part promoted by the religious elements in their beliefs.

Secondly, there is the problem of the Northern catholic community, which tends to be taken as a monolithic nationalist community. Apart from the fact that there has been significant support for the alliance party over the last dozen years or so among the catholic middle classes, there is also a further factor. Rose's (1971) survey, conducted in 1968 on the verge of 'the Troubles', found that as many as 30 per cent of the catholic population were 'explicitly prepared to endorse the constitution' of Northern Ireland, which then included the Stormont parliament. Catholics, however, did not vote for the Unionist Party. This indicates an acceptance of the state structures, or a recognition of a need for the separation of Northern Ireland from the remaining state structure of Ireland, but at the same time a continuing opposition to protestant–loyalist dominance. One can also add that there is something in the present policies of the SDLP which suggest a need to maintain a somewhat fragile unity in respect of the national question. Seamus Mallon, MP of the SDLP may have seen the Hillsborough agreement of 1985 as a step towards a united Ireland, but some of his co-politicians in the party would not share that view. Perhaps one should note the care with which the present leader of the party, John Hume, chooses his words over the future of the North. He is careful not to promote the theme of reunification, but talks of discovering 'a new way', setting up structures of co-operation in the North between the two Northern Communities, with the support of the Southern government. However, the policy is seen as ambiguous. Those who wish, both unionists and nationalists, can interpret it as the beginnings of absorption into an all-Ireland state. But those who wish for a basically Northern solution, even a dual national state with inputs both from the Republic and Britain, can interpret the policy as suiting their needs, and can thus find space on the ballot paper for the SDLP.

While in the last century there was still some attempt to embrace all people in Ireland as the Irish nation, now the nation subsists in the Irish catholic population. The ambiguity present in O'Connell has arguably aided catholic nationalists over the years in the justification of their policies. They have sometimes stressed religious elements why certain political changes were not possible, such as in legislation on public morality, schooling, marriage, divorce, and adoption. At other times, they have stressed the cultural, ethnic, historical, and geographical reasons why Ireland should be unified politically. At the same time, in theory they wish to accept that Ulster protestants form part of the Irish people while they actually do not accept them in practice. Indeed, until the ecumenical era of the 1960s, few catholics in Ireland were prepared to put before themselves some of the political religious issues which would have to be resolved if a united Ireland or accommodation to a separate political entity in Ulster were to become a possibility.

The other alliance in Ireland, the protestant–loyalist, does not thus far perceive itself as a nation. The view that they are one appears mainly in the writings of the British and Irish Communist Organization (1972; 1973). However, there can be little doubt that significant numbers of protestant loyalists see themselves as a distinct, almost ethnic group. Clearly, they can be seen to have a particular self-concept, a sense of what Weber termed social honour (Wallis, Bruce, and Taylor 1986). A group which perpetuates itself by endogamous sexual reproduction usually has some perception of ethnic identity. It tends to have an active or passive belief, whether articulated or not, that it is either superior to, or rightfully distinct from, other similar groups. The theoretical or articulated ground for this awareness is a set of abstract beliefs and moral dicta particular to the group. One of these beliefs may be racial superiority, but does not have to be. The distinctive feature of the beliefs is that they assert or imply the honour or prestige due to the group over and against another or other groups, and consequently motivate the group to achieve such honour on the field of social conflict. Of course, such beliefs can be religious.

It may be that in the protestant–loyalist case this ethnic tendency may have priority over the British component of identity or be in conflict with it. The contradiction can be viewed in two ways. Either there are two different groups among the loyalists, subscribing to two different loyalisms. Or there is really only one group, but sharing a dual identity, with now one dominating their consciousness, now the other. The truth of the matter probably lies somewhere in between. As Miller argues, there is a long tradition among the protestants of Ulster of viewing the British state ambiguously. It is clear that in the past many of them have considered loyalty to be a two-way process, a contract or covenant, and that the state could be a traitorous party as well as the people. The tradition proved of crucial importance in forming the basis for the strategy of the alliance from 1911 until British recognition of their separate political claim in 1914. If the tradition still exists today—and it appears that it very much does so—it implies that the United Kingdom does not have for many loyalists a natural character of statehood in the way the Southern state has for catholic nationalists. Covenant politics has come to the forefront again with the Hillsborough agreement of 1985 and shows the extent to which the populist politics of the Democratic Unionists is rooted in protestant–loyalist tradition.

This foundational belief gives meaning to the more popular belief in the right of the people to violence. Rather than such force being seen as a prerogative of the state, it is seen by many in the alliance to inhere in the people of Ulster who over and over again need to assert their autonomous rights. Increasingly it is seen as the only option left in the face of the continuing Anglo-Irish agreement. Continuing or worsening strife could solidify covenant beliefs into a separate Ulster identity absolutely dominant over the British component, and demanding its own independent state form. It is important to note that in 1912–14 the protestant–loyalist group saw their right to violence as prior to that of the official state. It is also important to note that they saw the taking of law into their own hands as temporary, and pending the recognition by that state of its mistaken attitude towards them and their right to a degree of autonomy.

It appears to be increasingly the case that there are two major traditions of identity perception among protestant loyalists. One tradition is the covenant one and is antagonistic to a straightforward nationalist sentiment. It is adhered to by increasing numbers who have become disillusioned with British identity and have seen Paisley's words about the traitorous nature of British governments become reality as protestant–loyalist domination in Ulster has been continuously eroded. These are probably more likely to vote for Paisley's democratic unionism and to come from either religious, fundamentalist backgrounds or populist working-class and lower middle-class backgrounds. The other tradition remains British identity and still would seem to dominate the values of the official unionists. It is likely that a good many protestant loyalists oscillate between the two and still have to come to their moment of decision, one that is likely to be forced on them by future events.[5]

The Monopoly of the Majority

The notion of the majority occurs frequently in Irish politics. In the South, protestant loyalists are accused of subverting the wishes of the majority in the island by not consenting to a united Ireland. In the North, protestant loyalists accuse catholic nationalists of not accepting the rights of the majority in the statelet to govern as it sees fit. In both cases, democracy is invoked. But it is the theory of democracy as the right of the majority to rule which is seen as central, rather than democracy as the preservation of minority rights. In the North, the belief comes via the British parliamentary tradition of the two-party system and its accompanying electoral process of 'first pass the post'. For protestant loyalists, the majority means themselves. In addition, protestant–loyalist politics has always been a zero-sum activity: one either has a monopoly of power or concedes it to the opposition. Because the opposition is the catholic nationalists, concession means the end of the statelet as such. The meaning of democracy shifts even further once it is interpreted within the terms of the Calvinist principle of the Godly society, where it is the lot of the just to assume power and to guide the citizens in the paths of righteousness. With this religious overtone in Northern Ireland, the belief in majority dominance has justified ignoring the rights of the catholic–nationalist minority within the Ulster statelet. From 1985, it has been continually invoked by Paisley and Peter Robinson in their attempt to subvert the Anglo-Irish accord.[6]

The Republican and Nationalist Components

It is important to note that the political party system in the Republic of Ireland is largely based on the divisions in the national-

---

[5] For a more extensive discussion of the two traditions in protestant loyalism, see Todd 1987; Fulton 1988.

[6] See e.g. Peter Robinson in a television debate with Seamus Mallon, 'A Week in Politics', Channel Four, 24 Oct. 1986.

popular consciousness which occurred at the time of the Irish civil war. To simplify a rather complex historical phenomenon, it can be said that the present political division between Fianna Fáil and Fine Gael originates in positions taken by opponents in that war. The civil war was not fought over the partition of the island as is popularly believed. It resulted from one group, the republicans, not accepting the oath of allegiance to the royal sovereign of the United Kingdom, a requirement of the constitution of 1922. Though the republicans lost the war, they eventually won their case with their successful introduction of the republican constitution of 1937. This shows the extent to which the republican ideal gained ground once independence was in place, furthering the ideological divide between protestant loyalism and catholic nationalism. Despite this development, the struggle between the two previously warring factions has continued, the original republicans developing a populist, nationalist party, and the former treaty party developing a concern with law and order, and moderation on the national question. But the fight did not prevent the fundamental beliefs in the nation and 'the historic integrity of the island of Ireland', as nationalist parties described it in their New Ireland Forum (1983–4: i. 28), from remaining basic to the perceptions of both parties.

The Gaelic-Irish Component

From the early years of the Southern state, cultural nationalism was deliberately developed by government means. There were deliberate attempts to develop elements of both high and popular culture in music, poetry, dance, and games. In particular, Gaelic games were promoted by the independent Gaelic Athletic Association. This association constitutionally still seeks to implement

> the organization of native pastimes and the promotion of athletic fitness as a means to create disciplined, self-reliant, national-minded manhood which takes conscious pride in its heritage of unrivalled pastimes and splendid cultural traditions as essential factors in the restoration of full and distinct nationhood.

By this means, the association believes the cause of national unity will be furthered 'until a complete nationhood is achieved' (Gaelic Athletic Association Rule Book (1970), quoted Hickey 1984: 40). Within this statement, which is rhetorical for some but deadly serious for others within the GAA, lies precisely the reason why the protestants of the North are so opposed to the development of

Gaelic games and seek to oppose their extension if and when they can.

The re-establishment of the Irish language as the first official language of the state became the national policy of the Irish government under de Valera and has only toned down in recent years. It was made a compulsory subject in the schools, an essential part of the Leaving Certificate, and it was necessary to pass an exam in the language to enter the civil service. A similar exam is still compulsory for anyone seeking to enter the teaching profession without the required Leaving Certificate pass, even if they are never to teach the subject. One immediately recognizes the active construction of the national ideal as promoted by principal currents in nineteenth century European nationalist thought (see Kedourie 1985). Another aspect of this form of cultural nationalism is that it has been strongly supported in the Irish national school system and Northern catholic school system by the clergy and religious orders. Not all the clergy and all the religious orders have been actively engaged in the pursuit of this ideal, but one or two orders have been; for example, the Christian Brothers are devoted to the ideals of a Christian and catholic education, and supporters of the concept of a nation dedicated to God and at the same time distinctly Irish. In making this comment, one is merely pointing out particular ideological characteristics in hard-working, deeply religious, and committed people. However, in the practical context of an Ireland divided between two power blocs, such a nationalist form is profoundly ethnic and exclusivist, whatever its believed aims and ideal content. There is an elective affinity between this cultural vision and the religious vision of the Roman catholic church prior to the second Vatican Council, in its exclusivist attitude towards other Christian churches, an attitude which is far from overcome at the level of popular Roman catholicism.

The Suppression of Class by the Dominant Beliefs

In the South, the marginality of the third main political party, the Irish Labour Party, is an indication of just how successful the articulation of capitalism to religious and national beliefs is. The South is not a classless society, but its class nature has been well hidden in consciousness and dominated by the procedures of the capitalist-democratic process. Despite the principal duality of Irish party political power, the bourgeois structure of the state has rarely been questioned, only the degree of state intervention in the bourgeois economy. Also, the clerical church has been particularly antipathetic to socialism in any form. It showed itself to have a horror of socialism already in the nineteenth century. But its opposition also had a base in the property-owning, rural peasantry, and middle-classes, who had in the nineteenth century paid a heavy price for their liberation from oppression and whose status became rooted in their property, land, and livestock. Their attitudes were shared by their priest-sons, who in their turn were supported by Rome, whose solid opposition to socialism until the 1960s was very clear (e.g. Pope Leo XIII, 1903). From 1911 to 1914, the Irish high clergy made successive attacks on the trade union movement and on socialism. Rumpf and Hepburn (1977) view the activity of the church during this period as quite decisive in limiting the appeal of socialism for the Irish subordinate classes until the 1950s.

The emergence of a new pro-capitalist party in the South from the end of 1985, the Progressive Democratic Party, might show the extent to which political divisions based on the treaty and within the nationalist component of hegemony may have become problematic, to be partially replaced by an even more explicit concern with increasing the popular wealth and prosperity. Indeed, in the Irish general election of 1987, progressive democrats took 14 seats, and 11.8 per cent of the total vote, which is sizeable for a young political party.[7] The decline of the labour element in the SDLP in the North, however, does indicate a trend to the style of politics in the Republic, that is towards the suppression of class politics and identity in line with the dominant ideology of the alliance. The national question as such still remains a central preoccupation for catholic nationalists. Because the nation is seen as only partially liberated, and particularly by the membership of the largest party Fianna Fáil, the animus of the national-popular consciousness is focused on this issue, to the detriment of class-based politics. There has, therefore, been little chance in the past for a political growth of class consciousness among subordinate groups.

What socialism there has been among the catholic–nationalist tradition has always tended to be allied to republicanism, especially in the period 1913 to 1930 (Rumpf and Hepburn 1977: 13). The trade union movement was a case in point. The labour movement

---

[7] For the 1987 election, see *Ireland Today*, 1987, No. 1035. For opinion polls in 1988, see Gallagher 1989: 153–4.

in Ireland was made up of two organizations. One of these was James Larkin's Irish Transport and General Workers' Union, the ITGWU. It came to have Connolly's support and was on the way to becoming thoroughly nationalist and republican under Connolly's influence. The other was the Irish Trades Union Congress. Before 1914, it became the Irish TUC and Labour Party and was neutral to the Easter rising of 1916. The ITGWU became a member of the ITUC in 1909, but had very distinct and separate policies to the moderate ITUC which sought to keep its links with Northern protestant workers. Socialists tended to be republicans, though only some republicans were socialist. The republicanism of Irish socialist nationalists was of course logical in so far as they interpreted imperialism as an enemy of the indigenous population and as an expropriator of the people. But it was thereby alien to Ulster protestant socialists. A duality of class interests within republicanism has remained, with forms of bourgeois and rural peasant-owner republicanism existing side by side with socialist republicanism. This combination still exists in the provisional movement especially in the South despite the officially socialist policies of Sinn Fein, as well as in the Fianna Fáil party.

In parallel with the Republic, class politics has flourished even less on the protestant–loyalist side, despite the periodic strength of protestant trade unionism (Patterson 1980). There did exist the small Northern Ireland Labour Party which disappeared after the fall of Stormont; but it had been based on industrial workers, particularly in the shipyards, and remained much smaller in size than its Southern counterpart. As Gibbon has noted, the cases of both Irish nationalism and Ulster unionism remain 'the two most spectacular class alliances in the political history of the British Isles' (1975: 3). The cause in both cases appears to be the strength of catholic–nationalist and protestant–loyalist popular consciousness and the way in which the cultural and material interests of the subordinate classes appear to be represented by the alliances themselves and their institutions. By providing core beliefs, and by reinforcing their separateness for both alliances, religious beliefs and institutions equally suppress class divisions, become embroiled in cementing alliances, and help retain the overall divisional structure of Ireland as a whole.

The Religious Component of Catholic Nationalism

There is nothing theoretically which hinders religion being an ingredient, or even predominating in the nationalist mix. Roman catholicism and protestantism constitute, in different ways and to varying degrees, nationalist and loyalist beliefs in Ireland. This is even official in the case of loyalism. In the case of Irish nationalism a specific relationship with catholicism has been formulated. Catholic identity is not essential for membership of the catholic–nationalist group, but consent to the understanding that catholic values will predominate in the public sphere is essential, for these religious values are viewed to be eminently suitable, natural, and self-evidently of high moral quality, safeguarding the fabric of state and public morality.

The link between Irish catholicism and Irish nationalism is deeply rooted. For centuries the Roman catholic church and faith were the bread of life to the subordinate classes in Ireland, deprived of their land, civil rights, and education. The church not only provided solace and comfort in those long years, but also a vigorous identity which enabled its people to hold up their heads amid the persecution and oppression. It provided intellectual and spiritual leadership and some alleviation from hunger and starvation. The clergy were the only source of education apart from the 'hedge school' teachers (Dowling 1968) and provided a significant moral and organizational resource.

Several aspects of catholicism became particularly important in the nineteenth century. As has been seen, partially as a result of Cardinal Cullen's nineteenth-century reforms, the church became more organized and developed in its numbers of clergy and religious. Its spiritual life underwent a considerable transformation. An important component of this religious renewal was its individualistic orientation; it was the individual rather than the group or its activity in the world which was the subject of this religious activity. Like all the religious renewal of catholicism in the post-Tridentine period, it was predominantly pietistic. The importance of conversion to Christ and subsequent pursuit of the way of perfection was the centre. It was strongly influenced by the development and multiplication of the religious congregational system developed by Ignatius of Loyala, founder of the Jesuits and of the famous spiritual exercises or retreat system. The exercises were specifically directed to obtaining this effect. A main real difference in the experience which marked it off from fundamentalist and some pentecostal forms of conversion experience was that the outcome was not the assurance of being saved, but rather a spirituality of trust in Christ based on repentance and the conviction that God forgives repentant sinners. In this respect, it paralleled some forms of evangelicalism, particularly methodism. The trust implied fidelity or commitment, together with an abandoning of the old ways of sin and indifference for a total dedication to God's will in Christ.

But if this was the ideal placed before the clergy and members of religious orders, the reality for the people was generally somewhat different. There can be little doubt, however, that a sense of God's presence and of the supernatural in general were long-standing characteristics of Irish popular religion, along with magical and semi-magical local religious practices. There can also be little doubt that the devotional revolution had a significant impact on the life of local catholic communities, especially as social life came to have a very strong relationship to the local parish and its priests. The transformations effected by the second Vatican Council of 1962–5 in liturgy, sacrament, scripture, and so on led to the accelerated decline of traditional rituals such as wakes and pilgrimages to local shrines. There has also been a significant decline in the cult of the saints, and many of the practices which characterized the devotional revolution of the mid-nineteenth century. This has been assisted by the continuing process of urbanization and delocalizing of religious experience. Local folk religion still continues in some forms. Its resistance is exemplified by the recent 'moving statues' devotional upsurge of 1985 in South-West Ireland, where statues of Mary were believed to have changed their posture, wept, and bled (Tobin 1985). National centres of devotion remain relatively strong—the shrine to Mary at Knock, the annual climbing of Croagh Patrick, pilgrimage and penance at Lough Derg.

In fact, whereas the parallel protestant movement of evangelical revival was for all people, and the issue tended to be decisive, catholic spirituality diverged on both points. The principal one was that conversion and growth were particularly directed at the clerical-religious élite, where the annual retreat system formed the annual or biennial centre of the individual spiritual life. The retreat consisted, and still does, of anything from five to thirty days of spiritual exercises, usually in an atmosphere of silence and contemplation apart from listening to the preacher when there is one, but with additional features of both vocal and silent prayer. The laity had a pale reflection of this programme in the parish mission, designed to convert the laity or at least bring them back to regular church practices. Morality was for the laity, whose life was dominated by the battle against mortal sin, and who therefore lived under the threat of hell and were always at risk. Spirituality was the lot of the clergy and the religious, who were judged to have left this crude level of human religious existence behind, through conversion and a subsequent, progressive spiritual change. They were seen to succeed in living a sexually pure life as part of this. There was some extension of the pietistic ideal. Sodalities, such as the Legion of Mary, Opus Dei, and Christian Life communities, have partially extended this form of commitment to some, particularly more middle-class laity, and continue to have an important role in activating laity for what are judged to be religious goals both personally and socially. But generally speaking the ideal has always tended to accentuate the gap between the clerical world view and the lay world view within catholicism. Even though the gap between clerical and lay religious intellectuals has closed, with clergy being left behind in some areas, the clerics remain the true *cognoscenti* in religious matters, and are expected to be so by the laity. This has always added to the clergy's spiritual authority and status, and has tended to merge with the authority claimed by the clergy in matters of faith and morals, with the high clergy deciding what constitutes matters of faith and morals.

The attitude towards authority within the church partnered the perception of the spiritual life. The clergy were and still are in a special position, holding power over the laity because of it. This is one of the special characteristics of catholicism, and is found to a much lesser extent in non-episcopal traditions. Failure to conform to directives of priests and bishops were seen to break the very fabric of communion and to allow sin to enter in. This remains particularly so today where an objectivist stance on morality prevails in a national or local culture anywhere in the catholic world. From within this perspective bishops and clergy lay down rules for the laity to follow in any given situation and the teaching of the church is seen as absolutely clear and devoid of problematic. Rules are enforced rather than principles declared.

In terms of the Irish catholic–nationalist context today, pietism and authoritarianism have tended to structure the religion of the people to a significant extent, though the numbers subscribing to its world view appear to be declining. The clergy are seen as above criticism in their religious statements, and such criticism can cause considerable distress to many people. In his analysis of the popular culture which appeared among the promoters of the Pro-Life Campaign, set up to achieve a constitutional ban on abortion in the Republic in 1983, O'Carroll pin-points certain characteristics, which can be abbreviated here: a monolithic and absolute view of the world, with its accompanying intolerance, derived in part from the direct consultation of clerics and politicians on public moral issues and the subsequent failure to develop an ethos of public debate; a localized belief system, rooted in family and communal authority and issuing in a spirit of absolute conformity; sexual prudery, a product partly of the inheritance problem; and the development of acute anxiety when such beliefs—inhering partly as they do in their practice and shaping of society—are threatened. The absoluteness of the vision excludes all alternatives, and requires external validation. This can be provided by the magic power or words, such as inscribing prohibitions into the nation's constitution, and by manifestations of the divine in terms of either retribution for failing to uphold the truth, or grace for upholding the truth, with rewards 'such as oil finds and victories in the World Championships'. O'Carroll sees such positions exemplified by the protagonists' refusal to accept that certain therapeutic abortions already permitted in Ireland were actually abortions and by their predictions of social chaos should they fail in their fight: 'Hence phrases such as "the opening of the floodgates", "the thin end of the wedge", "the slippery slope", "the permissive society", and "the abortion mentality"' (1983: 12).

It is important to point out that these expressions which interpret Irish social reality and events are also used by the high clergy. They are thus not simply a mentality derived from popular religion but from a traditional Roman catholicism which held sway in catholic Europe from the post-Reformation period and remained unchallenged until the 1960s. As will be seen in Chapter 5, understanding this religious social consciousness requires some grasp of the traditional catholic teaching on the natural order and the good society, and how the nation is to respect the divine order established by God. An example of this can be taken from the recent contraception controversy in the Republic, which began in the 1960s. At that time, the Roman catholic archbishop of Dublin intervened in a pastoral letter in the following revealing terms:

> If they who are elected to legislate for our society should unfortunately decide to pass a disastrous measure of legislation that will allow the public promotion of contraception and an access hitherto unlawful to the means of contraception, they ought to know clearly the meaning of their action, when it is judged *by the norms of objective morality* and the certain consequences of such a law ... It may well come to pass that ... legislation could be enacted that will offend *the objective moral law*. Such a measure would be an insult to our Faith; it would without question prove to be *gravely damaging to morality, private and public; it would be, and would remain, a curse upon our country*. (Abp. McQuaid of Dublin, letter to all the churches of the archdiocese, Mar. 1971, quoted Wright 1973: 224; italics added)

It is the specific effect this religious form has in Ireland which is under scrutiny. But the form as a whole should be recognized as inimical to protestants, especially when pursued in the arena of politics. This will be considered in the remaining chapters.

The Protestant Element of Protestant–Loyalist Beliefs

The dominant religious element in protestant–loyalist beliefs is a form of pan-protestantism. It has so far overcome at the political level the doctrinal and organizational divisions between protestants. It is based on the theme of protestant liberty, which is consciously contrasted with catholic authoritarianism. There are, however, many strands within this protestantism and they are frequently conflicting.

One has to remember that well into the nineteenth century British identity was predicated first of all of the English. It also tended primarily to be Anglican and only secondarily reformed. Even when non-Anglican protestants had become accepted as citizens in England and Ireland towards the end of the eighteenth century as they already were in Scotland and Wales, English and Irish catholics remained politically suspect. They remained so throughout the nineteenth century and were not normally trusted with public office. Also, Irish catholics were represented in the dominant culture as an inferior race, ridiculed, and considered incompetent and beastly (Curtis, Jr. 1968). The idea of defending a protestant heritage is not solely an Ulster phenomenon but rooted in the British tradition. Whereas in Britain, with the growth of suffrage, catholics could never be perceived as a political threat because of the smallness of their numbers, in Ireland, with the growth of the home rule movement and its accompanying nationalism, they could only be perceived as politically, religiously, and nationally subversive and suspicion of them remained.

The issues of freedom of conscience and freedom from Roman hegemony lie at the centre of Northern Ireland fundamentalism and are meshed in with its evangelical tenets. They are at their clearest in the teachings of the appropriately named free presbyterians:

Liberty is the very essence of Bible Protestantism. Tyranny is the very essence of Popery. Where Protestantism flourishes Liberty flames. Where Popery reigns Tyranny rules.

As Liberty and Tyranny have no common meeting place, so Protestantism and Popery cannot be reconciled . . . Popery is tyrannical in every sphere of life . . . Protestantism, at a stroke, cuts down all the shackles of superstition and priestcraft. The soul is free to commune with its Maker. There is one mediator between God and man—the man Jesus Christ. Protestantism hands out an unemployment card to every Papist priest . . . The Protestant's home is his castle. He brooks no interference from Pope, priest, pastor, preacher or prelate . . . No one can control the Protestant's education or the books which he shall read. He is free born and trembles not at priestly threats or the Papal curse . . . At the ballot box the protestant exercises complete liberty . . . (Ian Paisley, *Protestant Telegraph*, editorial, 28 May 1966, quoted Wright 1973: 224)

Despite this common, oppositional identity, protestants in Northern Ireland retained institutional, class, and doctrinal divisions. Presbyterians retained a consciousness of their Scottish origins and came to have a significant presence in the local economy and finance institutions, predominating in the North and East. Church of Ireland members, largely of English origin though less conscious of it, predominated in the south of Ulster and still constitute the majority among the protestant farming community. The small denominations and sects are province-wide, but are particularly strong in Belfast. They are more fundamentalist both in religion and in politics—though some eschew political activity—and are frequently more politically volatile. Both Church of Ireland and presbyterian political ethics in the North are supportive of the union and the forces of law and order, whereas the sects have a more flexible respect for state authority.

One is not talking about a lack of breadth in theologies and points of view within presbyterianism and the Church of Ireland. Both have their liberals, religious intellectuals, and deeply committed ecumenists. There is a tradition of strong social consciousness in the Church of Ireland, evidenced by its building of hospitals, the support of some of its leadership for the locked-out Dublin workers in 1913 and their activities in other works of mercy without a proselytizing aim. A number of higher clergy in Ulster and many of its lay intellectuals stand a long way off from protestant–loyalist politics and are in fact politically dissociated from them. In the South the church accepts the status of the independent republic and is loyal to it. But the bulk of its laity and Northern clergy, and probably more so its industrial and farming classes, are low church and evangelicals, the more protestant and anti-catholic end of the Anglican synthesis. One must also remember in this respect that a key feature of Church of Ireland organization is the local select vestry, which involves the laity in debate and discussion with clergy. When placed alongside the house of laity in the synod of the Church of Ireland, the power of the laity in church decision-making, including statements on political ethics and positions on public local issues, is quite considerable.

A similar statement can be made of the presbyterian community, which is almost wholly Northern, and whose main Southern presbytery is in the counties of old Ulster which remain in the Republic—Donegal, Cavan, and Monaghan. The organization of presbyterianism rests strongly upon the local community, with its division of authority between minister and elders. The elders are individuals of good-standing drawn from the local laity, who both appoint their minister and have considerable control over his or her activities—there is at present one woman minister with a congregation in Northern Ireland. The sentiments among the majority of Northern lay presbyterians are biased towards those elements in the *Westminster Confessional* which are antagonistic to Roman catholic practice and belief: the evils of the papacy and priesthood and the suppression of Christian liberty, the evils of the mass and devotions to the saints, all of which combine to turn Christians to superstition and idolatry and away from almighty God and true salvation in Christ. An example of the strength of this animus occurred in 1982–3 in the celebrated case of the Revd David Armstrong, censured by the elders of his congregation for entering the Roman catholic church across the road to offer its worshippers Christmas greetings, and eventually forced to leave his ministry (Armstrong and Saunders 1985). Though the case is complex and the minister has joined the Anglican communion, the elements of the case at the time are quite significant. Ministers, both presbyterian and Church of Ireland, suffer similar pressures in many other local communities, except that they often stay silent rather than risk conflict, personal hardship, or violence. An added fear is losing church members to sects or the free presbyterians.

Protestantism has a deeper significance in Ulster than it has had in Britain for over two hundred years. The very phrase 'the Ulster people' refers to those who originally set up the true faith in Ireland and colonized the land accordingly. This religion of the people has some similarity to Calvin's theology of the city of God on earth, though in a more modern and less rigorous setting than the one which Calvin tried to realize in his lifetime in the city of Geneva:

Civil government is designed, as long as we live in this world, to cherish and support the external worship of God; to preserve the pure doctrine of religion, to defend the constitutions of the church, to regulate our lives in a manner requisite for the society of man, to form our manners to civil justice, to promote our concord with each other, and to establish general peace and tranquility . . . (Wright 1973: 225, quoting Calvin 1960: 46)

Calvin viewed his interpretation of the function of the state as the only Christian one. This mode of political religious action no longer starts out from a universal centre and figure, such as the papacy, but rather from the national or local church within the state, whose 'magistrates'—Calvin's term for lay political leaders—are ideally Christians of moral rectitude, who perform this duty as one ordained by God. Wright is correct to see Paisley's interpretation of liberty as a development of this view. Both reject papal centralization and papal authority as a means for discerning just government. Wright is also correct to distinguish between a general interpretation of 'salvation by grace'—the teaching of the reformers—and its fundamentalist variant, which he terms 'salvation by grace through faith'. Perhaps 'salvation by grace, experienced in listening to the preaching or reading of the biblical word, and in affirming the literal interpretation of that word' might be a better if clumsier phrase. In its effort to avoid all church mediation of the revealed word of God, the fundamentalist version relies on the literal interpretation of the Bible, but one which imposes a contemporary Western culture on to the different perceptions of a past culture. It is this characteristic which renders fundamentalist religion open to large numbers of fissures in communities, with preachers or sections of congregations shaking the dust from their feet and moving off to build or buy another church. Fundamentalism is radically anti-catholic in its refusal to accept any form of priestly mediation.

Hickey (1984) and Wright (1973) grasp the point that both fundamentalism and Calvinism have monopoly tendencies in the public sphere. The consequent interpretation of liberty is anti-libertarian and anti-pluralist. The implementation of liberty can stretch to keeping places of recreation closed on Sundays, upholding the rights of preachers to speak in public places even if it causes an affray, and opposing the development of catholic schools because their teachers do not communicate the Bible without priestly mediation.

At the level of the unchurched popular consciousness, particularly in the urban protestant working classes, liberty may have slightly different significances. It may also mean freedom to purchase 'Playboy' magazine, to indulge in the vulgarities of verbal and visual displays of the sexual and sexist types, though at a much lower level than those current in Scotland or England at soccer matches and in public houses. But there remains a strong respect for the religious leadership of the protestant–loyalist bloc. There is a resonance of protestant beliefs in their unchurched experience, their rough, straight-from-the-shoulder speech, their interpretation of history and tradition, and solidarity at the economic level based on their dependency upon the leadership of the group as a whole for their work and welfare. The working classes are strongly bound into the total if ambiguous ideology of protestant loyalism.

With the emergence of the statelet in 1922, covenant politics was subordinated to the Stormont alliance only to re-emerge after Stormont's collapse in 1972. In 1971, a new covenant was signed by over half a million men and women, committing protestants to a renewal of the covenant political ideal. Miller (1978) suggests that the contractarian element was no longer ever so solid. Be that as it may, it may turn out that the preparedness not to acquiesce in any all-Ireland solution is decidedly stronger than most commentators until recently were prepared to accept, except for the work of Bew, Gibbon, and Patterson.

Thus, protestant loyalists fight incorporation into a united Ireland for the reason that their perception of protestantism and their values of polity contain a powerful rejection of catholic monopoly, which they identify with catholicism *tout court*. Their deeply held view on this matter is bound up with their interpretation of political realities. The South is seen as dominated by the Roman church. They point to the laws against divorce, abortion, and contraception, the power of the church in the education sector and past evidence of a hot-line between bishops and politicians as the proof of their contention. As will be seen in Chapter 5, this interpretation is not without foundation. The ethnic identity or contractarian ideology of the Ulster loyalist is clear in its origins, motivated by powerful economic, ethical, and religious convictions, all of which are interrelated, mutually supporting, and validated by the existence of a dominating catholic–nationalist bloc in the Republic of Ireland. The sense of solidarity has developed and hardened over the years. The popular slogan 'This we will maintain', that is the union with Britain, can be equally applied to the religious and moral values the loyalists pride so much and which they see as embodied in the protestant–loyalist statelet, with its appropriate state apparatus of coercion and alternative paramilitary units. For this reason, they feel that the siege of Londonderry has never been lifted and they are prepared to die to defend that heritage. What the protestant loyalists in effect seek to maintain is protestant–loyalist domination within a flexible territory, one which they feel justified in calling their land and ruling according to their conscience.

The Key Mythical Structures of Protestant–Loyalist Popular Religion

It could be argued that the myths of Ulster protestantism and the institution of the Orange order take the place of the more centralized clerical organization of Roman catholicism in providing some element of overall religious unity among protestant loyalists. By myth is meant here what has generally come to be accepted within sociology and social anthropology since the work of Lévi-Strauss: an account of the origins of a society or of particular crucial events in its life, which unite the cosmos to the social structure by actively shaping everyday life perceptions. The historical consciousness of Ulster protestants in this sphere is also a faith consciousness. The salvation history of the Bible is coupled with what is seen as a significant period in the history of the protestants of Ireland.

The key event in the history of Ireland which appears to encapsulate Ulster protestant faith in God through Christ and the destiny of Ulster protestants as a people, is the war of 1689–91. William of Orange is seen to have worked the decisive victory at the Battle of the Boyne in 1690. However, that myth is not quite the one which appears to have penetrated everyday life, perhaps because the place of the victory is located within the borders of the Southern state. The most powerful myth affecting everyday life narrates the less decisive defence of the city of Londonderry in 1689. The protestant refugees, inhabitants, and garrison of the city, which was built deliberately with a wall lest a catholic rebellion should ever occur, held the city for fifteen weeks against a poorly equipped catholic Irish army led by James II, the exiled king. The besieged were led by a soldier and a clergyman. A further particularity stressed in the protestant–loyalist account is the initiative and bravado of the local protestant apprentice boys who defied the city's governor, Robert Lundy. He had ordered the opening of the city gates and capitulation. The boys barricaded the gates and mounted the city walls, a move probably as much a result of a popular rebellion against Lundy's action as a defiant gesture. Once the blockade of the river leading into the city was broken by English ships, James and his besiegers lost heart and abandoned the siege.

The mythical value of the siege for the construction of protestant–loyalist hegemony should not be underrated. In his highly condensed and powerful analysis, Buckley (1984) details how the siege becomes interpreted by today's protestant loyalists according to a key paradigm mediating their experience of the world. They may be talking about membership of Ulster's protestant secret societies, particularly the Orange order, the Royal Black Preceptory, and the Apprentice Boys' Club; or about religion—being saved, being a church member, or even a non-practising protestant; or what appear to them to be key ethical issues such as drink, tobacco, and money; or they may simply be describing in ordinary everyday language life in the family, on the farm, and in the village. But in all these instances, similar categories of differentiation occur—the insiders and the outsiders, the religious and the rough, the parent and the child, the female and the male. At one point, the protestants who are involved in violence are rough as opposed to religious, especially if they drink and smoke and do not follow the norms of thrift. Shift the context and the same men become insiders as opposed to outsiders such as catholics and, presumably, even castle catholics. The rough insiders are on the walls, defending God's true believers from Romanism and immorality. As Buckley points out, it is also the rough who penetrate the virgin city—the British fleet, breaking the boom across the river to bring food supplies to the stricken protestants within and ensuring the city's survival and the faithful people's triumph. One can recognize here, with Lévi-Strauss (1968), the importance of ambiguous mediators in relating the opposites within one's cultural universe and one can add the Apprentice Boys to Lévi-Strauss's shortlist of tricksters and twins in this role. Despite Buckley's reluctance to play up the siege of Derry as a key myth and his preference for general categories of interpretation derived from various historical experiences, it does seem that the siege has particular significance. Those who still seek compromise with catholic nationalists today are still known to the erudite leadership of the DUP as Lundies.

One should add that Buckley takes the Christian myth for granted, as backdrop to the discourse. But one must remember that the divergent reading of the Christian myth by Reformation and Counter-Reformation is at the heart of the religious as opposed to the rough interpretation of the conflict. The importance of the protestant variant should not be underestimated given the extent of religious practice and the power of religious men within the protestant–loyalist bloc and the grave importance of conceiving of one's position as absolutely irreproachable, depending as it does on a divine source. At the heart, or within the innermost wall of the protestant–loyalist alliance, is the man of God—not woman, not rough, and not immoral—be he soldier, clergyman, businessman, or farmer. After the account of creation and redemption, which despite the variance among intellectual accounts in the past probably has slightly less relevance as a divider at the popular level, the two main variants begin (for popular accounts, see Harris 1972; Leyton 1975). In the Roman catholic account, Christ appoints Peter to lead the church he has founded. In the other, the church is founded but taken over early on by the Popes in Rome and by priests who use magic and ritual to gain control of God's people, thus profaning his divine word. For the one version, the Reformation is the catastrophe by which half Christendom defects from the church, the church itself remaining intact and faithful to Peter. This version still predominates among Roman catholic laity in most countries. For the popular protestant version, one which is still shared probably by a majority of clergymen within the protestant denominations of the North, the church re-emerges after centuries of misguidance only with the Reformation. The protestant version has variants as to how pure this church is or whether it remains sinful, but the evangelical version which is the basic one in the North of Ireland opts absolutely for 'Jesus Saves'; the community, no longer medium of salvation, tends to take on a visible, earthly role, an occasion of grace and a support for religious 'this-worldly' activity. Rome is seen to continue on its headlong path to misery and sin, a symbol of the rejection of Christ for the sake of human idols. However, the earthly role of the community, be it protestant or catholic, is not seen as secondary to the religious message, but as either in continuity with or against it. The Calvinist vision of an earthly city in which pastor and politician work in harmony to exclude evil from the lives of all those within the city walls remains a religious outlook for a significant number of protestant loyalists. This is a totalizing view of society, and implies a certain anti-pluralism: freedom of belief and action within certain parameters, which are to be decided either by those who appear righteous in the sight of God or by those who, at least, conform in their lawmaking to the advice given to them by the righteous. The very drift of the account of Christian myth and religious political ethics has certain parallels of structure with the siege of Londonderry. These should be noted, before one views the siege as baseline myth for the interpretation of everyday life.

Religion and Violence

It is frequently denied that there is a relationship between religion and violence in Ireland. Here it will be argued that an indirect, and occasionally a direct, relationship does exist. Before applying the general analysis to Ireland, and in order to get to the heart of the matter, it is first necessary to understand the close relationship between beliefs and violence within the context of the state in general.

It is the state which properly exercises the use of force in order to maintain social control, whereas it is in civil society—the family, school, work-place, church, and other non-state institutions—that hegemony operates. However, the state also performs certain hegemonic functions too, justifying its own existence and authority in the rhetoric of the courts and in its celebration of the nation. The civil sphere can also be an arena for coercion, through the activities of vigilantes, private armies, and mobs, irrespective of individual criminal activity. Even so, force is, in the ideal state, primarily an exclusive instrument of the state, though even here force diffuses outwards from the state to civil society and to sections of its population and, as such, is intimately connected with hegemony. This is already true at the level of the institutions of law. These express the beliefs of law and order, and justice and freedom, hegemonic beliefs necessary for both citizen and state in any society aspiring to some model of equality. This ideology of the state is coercively implemented in the policing, arresting, trying, convicting, and imprisoning of those found to default. But this coercive role is partly performed by citizens too. Sections of the population assist the police in the performance of their duties not just by taking on gunmen or robbers but also by making telephone calls to the police. They thus form an extension of the coercive dimension of the state. They consider such actions as their duty as citizens. Put in another way, force is underpinned by dominant values extensively shared by the population and which support the operation of the state. Citizens then act as witnesses in the courts and finally may compose the jury—the real judge of guilt and innocence.

In all capitalist democracies there are two sources for the legitimacy of the state, namely statehood itself and the ideology specific to the state in question. It will be argued that when the second source is threatened as well as the first, the power to coerce can devolve on the civil sphere in a substantive way. The conclusion to the argument will be that there is a specific form of violence of a political kind outside of state control which cannot be reduced to crime. The specificity of this form of violence or coercion has to be recognized, however one feels about the terrible consequences of it for the innocent. This form of political violence requires political answers and not just an anti-criminal, police-oriented, coercive response. But also because such violence can have some hegemonic justification—derived in Ireland from the threat of the alternative alliance—the beliefs too must be scrutinized for their contribution, religious beliefs as well.

We can look at the twin sources for the state's legitimacy in the context of the Republic of Ireland. First there is the fact of statehood, with the implied acceptance of law and order by the people. Second, there is the unifying dominant beliefs of Irish catholic nationalism. The distinct importance of the two sources appears in the way in which state and society in the republic treat the provisional movement. PIRA is a proscribed organization and there is also a law preventing the national radio and television, RTE, from giving air time to members of the movement. The intention is to impede the divulgation of its aims. At the same time, despite the popular support for many of the activities of the provisionals north of the border, there is almost no popular support for terrorist activities against the established government in the South. The legal position of the IRA is thus indicative of the inviolability of the statehood of the Republic in the national–popular consciousness. An anti-state force such as the provisionals, who do not recognize the statehood of the twenty-six county state, cannot achieve support in the South for this essential policy. The power of the element of statehood among the dominant beliefs appears to have been sufficient to neutralize the provisional movement in the South, even though it was severely tested by the event of Bloody Sunday in 1972, when thirteen demonstrators were killed by British soldiers in Derry, and by the provisional hunger strikes of 1981. Any further attempt to violate the principle of statehood by the new Northern-led leadership, it could be presumed, is unlikely to succeed—indeed, in certain circumstances could result in a reduction of support for their Northern campaigns.

Secondly, and in contrast with this respect for the state's monopoly of violence, because the provisionals seek to destroy the Northern dominance of protestant loyalists with a view to restoring 'the historic integrity of Ireland', there is and will remain substantial support from among the Southern population for the violence the provisionals actually commit in the North. The belief in the integrity of Ireland is central to the unifying dominant beliefs of the alliance and is clearly distinct from the specificity of statehood. The dual source of the state's role in hegemony helps us understand the limitations of the state in respect of the alliance whose power it also represents. This is why collections by the provisionals in public houses were having partial success in the

1970s without too much interference by the Garda, the Republic's police force. The collections indicated a wider hidden culture of support for 'the lads' who were at the forefront of the battle to restore national unity, in addition to the much smaller party political support they enjoyed in the South.

The Roman catholic church, in episcopal statements, unequivocally condemns violence North or South. As seen in Chapter 3, this was also true in the nineteenth century, despite lower clergy support particularly for the Fenian movement. Cardinal Cullen succeeded in getting three of the four religious provinces of Ireland to deny the sacraments to the Fenians, and only Archbishop McHale of the province of Tuam failed to toe the line. Generally speaking, there is evidence that the church has always tended to support the hegemony of the state *qua* state. It supported the British state for most of the nineteenth century in Ireland, then supported constitutional politics for home rule; and when independence came in 1921, it supported the pro-treaty faction in Southern Ireland and opposed the civil war because that faction was the state.

It would appear that the popular religious consciousness makes a distinction between violence against the Southern state, whose statehood cannot be questioned, and violence for liberation in the North. Though the extent to which such a consciousness exists is not clear, one aspect of it is beyond dispute: 35 per cent of McGreil's random sample of Dubliners and 46 per cent of the males in the sample, including a spread from the younger age groups and the more educated, supported the view that 'the use of violence, while regrettable, has been necessary for the achievement of non-Unionist rights' (Mác Gréil 1977: 387). In any event, there is likely to be significant variation in views as to how far one should condemn or condone this liberationist violence in the North.

One can now see how the preliminary data dealt with in Chapter 1 on good and bad religiosity among Roman catholics is indicative of a long-standing historical contradiction. Irish catholics find themselves in an ambiguous relationship to Irish nationalism. Having become the political expression not just of the state appropriate to Ireland but of Irish catholics as well, the Irish nationalist tradition contained and still contains within its confines a culture of violence antithetical to the church's traditional teaching. Catholic nationalists who support or condone the killing of protestant loyalists do so on the grounds that members of the

Northern police force and part-time army are either agents of British imperialism or maintaining alien protestant–loyalist power in a part of Ireland. Some more theologically articulate members of the provisional movement justify violence on traditional Roman catholic ethical grounds, the theory of the just war against the unjust oppressor. For this faction, the term 'just war' has secular as well as religious appeal, an ambiguity which permits them to communicate their cause in direct fashion to left-wing British and right-wing American groups at one and the same time.

This culture of violence is complex, and needs wider and deeper research. One is not simply alluding to the tradition of Padraig Pearse, one of the leaders of the Easter rising of 1916, who paralleled national violence and bloodshed to the salvation effected by the blood offering of Christ. That tradition was frequently eschewed by later republicans. The broader tradition is a typically nationalist one, seeing national liberation through war as honourable and singularly justified. To a degree it is widespread among many not just in the provisional movement but also in the Fianna Fáil party, and in general among those who see their history as one of liberation through blood. The statue in the market square at Dundalk, a Provisional Sinn Fein memorial, of the Irish liberator rising up behind the phoenix, is an image with much wider appeal. The Easter rising of 1916 as an act of national liberation, unquestioned in the dominant catholic–nationalist culture from 1922 until well into the 1960s and 1970s, still holds a cherished place in the hearts of the majority of the people. It is possible for such sentiments of approval of this past to coexist with abhorrence for most current acts of violence.

But the political significance of this culture is that where opinion counts and where the catholic–nationalist remnant actually experiences the coercive power of protestant loyalists and the British army in the Northern statelet, there violence has all the more support. When it does, it tends to be a reaction to perceived injustice, such as internment without trial, or the conviction of a son by a sole judge in a trial held in total secrecy and on the evidence of unseen witnesses, or a simple case of one's house being badly mauled by careless soldiers searching for arms.[8] This reactive violence is thus justified by its subjects and the justification subsists

---

[8] The writer came across alleged cases of heavy-handed house searches in Belfast in the mid-1970s.

in consciousness along with their Roman catholic profession of faith and identity.

The provisional IRA's commitment to violence against the British and against the protestant–loyalist alliance, which the provisionals rhetorically and conveniently subsume under the term 'the British', is frequently assumed to be based on either Marxist or nationalist principles and in both cases to be secularist or areligious. Yet some provisional leaders and rank and file members are committed to their Roman catholic religious belief and practice. These beliefs can in fact contribute to the motivation of their struggle. As Gramsci points out, the religion of the clergy is not necessarily the religion of this or that social group of laity, whose religious interpretation is also related to their own concrete world of experience. Catholic morality approves of the view that to repel an aggressor is to engage in a just war. It just so happens that Irish national–popular consciousness interprets the presence of the British state in Ireland as a form of aggression and hence feel justified in repelling the British from Ireland. Catholic morality also approves of a proportionate degree of violence to overthrow tyranny. Many Irish catholic nationalists also consider the British and their Orange allies to have established tyrannical government in a part of Ireland. Therefore they find it just to overthrow the Northern statelet by whatever degree of force proves necessary.

Of course, some beliefs in violence are unrelated to catholic moral teaching. Some Irish nationalists hold that it is just to unite the nation by force—a typical view of secular nationalism throughout the nineteenth and twentieth centuries in Europe. But traditional catholic moral doctrine would oppose this on the grounds of the legitimacy of the state *qua* state.

Doubtless some protestant paramilitaries—though there have been fewer religious persons among them than in the case of the provisionals—hold to similar religious understandings of the necessity of violence under certain circumstances. Covenant theology itself empowers the people to take action against the state in certain and specific circumstances, and Ian Paisley and many of his supporters both within and without the Orange order have frequently used bully-boy tactics, and approved paramilitary processions, marching, and drilling. From Carson to Paisley one finds the argument that it is legitimate to fight to regain one's fundamental freedom if the sovereign with whom one has covenanted one's allegiance betrays that allegiance and conducts one into slavery. The argument is clearly inspired by a religious motivation and shows Ulster protestantism to be a vigorous form of political religion, approving whatever defensive measures are necessary to avoid submission to Rome rule. Ulster protestantism is not contained within churches but spills out on to the streets.

Religion is also involved in the violence indirectly. Indeed, this is the most significant aspect of the role of religion in the divisions and conflicts in Ireland and goes to the heart of the matter. In the remaining chapters, it is argued that religion in Ireland basically solidifies the opposing alliances in an exclusive way and precludes a common state form, thus providing the structure of violence which such divisions entail. This will be shown to be particularly the case in the next chapter on the relationship between catholicism and the Irish constitution.

CONCLUSION

It would be true to say that catholic nationalism was at its most formidable in the years between independence and the rise of pragmatic politics in the Republic under Sean Lemass in the late 1950s. In that period of the early development of the state, the nationalist ideal was almost totally unquestioned. But now, on the basis of a number of well-documented opinion polls, W. Harvey Cox (1985) argues that the majority in the Republic of Ireland favouring reunification is barely two to one, and that in the island as a whole it is three to two. What is more, Cox suggests that support in the South is at best lukewarm. None the less, the characterization of catholic–nationalist ideology I have just documented would seem to fly in the face of this sort of evidence.

The general data presented by Cox are undeniable, but interpretation of the data rests on certain assumptions, principally that people act on an individualistic basis rather than as groups, and according to attitudes expressed in the quiet of their homes rather than in response to major events. One remembers the substantial surge in Provisional support in the North on the occasion of the hunger strikes of 1981, and the impact they had on Northern catholic–nationalist consciousness, one which Cox himself acknowledges. One also remembers the effect 'Bloody Sunday' had on Southern attitudes, resulting in the burning of the British Embassy in Dublin, in 1972. Though opinion in the South gradually softened, there can be little doubt that the obverse would have been the case had the British government not changed tack and demurred at an openly repressive strategy. Public opinion responds to the conjunction of culture and event and solidifies either old or burgeoning values as might be the case.

What has been attempted in this chapter has been the clarification that *so far* the traditional values of catholic nationalism are still dominant among the Roman catholics of Ireland, even if they are severely contested by a significant minority from among them and lukewarmly subscribed to by a further significant minority. There is as yet no shift in the overall direction of the group, which remains dominated in terms of political action by its monopoly catholic and nationalist constituencies. Cox sees the growth of modernity and pragmatism as the likely direction for the group, with the consequent and gradual dissolution of catholic nationalism. That is indeed a possibility. But more likely is the possibility that it will continue to respond to events, in the way it has done in the past—and there is no guarantee that the evolution Cox envisages will be any more natural than the modernization of Iran seemed to be in the 1970s. The comparison is a little strong, but only with the intention of making an important point. In 1985 Cox's paper was timely and asked for shifts in American Irish and British labour politics in a wholly correct way. But the optimism for change in the Republic of Ireland remains somewhat unqualified.

5
Religion and Law in the Southern State

Shoring up Monopoly Catholicism

INTRODUCTION

One of the key institutions which embodies the values of dominant alliances in states is the institution of law. It is important particularly because it draws together both coercion and hegemony in the state. The dominant religion can also be mediated through law. In fact, in the Republic of Ireland, catholicism is doubly important. Not only does it form a part of the dominant beliefs of catholic nationalists, but their state form gives catholic social teaching coercive and hegemonic support. Catholicism is present in everyday life through state law, as well as through authoritative statements by clergy and through the national–popular consciousness. The clearest example of law as mediator of political religion in Ireland is the Irish constitution of 1937 and its subsequent interpretation. Part of this interpretation has been by politicians and clerics as well as by judges. The interaction of these powerful figures in the life of the Republic has been significant for the preservation and development of the catholic–nationalist world view. Religious leaders have developed ideologies of natural law and the common good to reinforce the Roman catholic element. Sections of the catholic–nationalist population have also combined their own popular nationalism and religion with aspects of the clerical interpretation already invested in the law, particularly in the anti-abortion movement of the early 1980s.

Other areas where political religion helps to solidify one or both of the alliances are separate education and catholic-protestant marriages. These will be dealt with in Chapters 6 and 7. The three areas can be described as ones of political–religious mediation occurring at different levels of social process. They help solidify state, church, and common people of varying classes into two opposing blocs from whose grasp it is difficult to escape.

RELIGION AND LAW, 1922–1972

The Irish Constitution and Roman Catholic Teaching

Sown into the constitutional fabric of the Republic of Ireland is a local and time-bound Roman catholic ethos which has changed perhaps in only minor ways since the founding of the state. The Irish state apparatus is certainly secular: priests have no role in it. The Roman catholic church is financed out of voluntary contributions. The state has no role in the appointment of bishops and the juridical processes of church and state are kept entirely separate, though it will be seen that a number of administrative procedures reflect a direct link. We will see that this mode of constructing monopoly catholicism is alien to the Concordat period of Roman catholic church–state relations recognizable in arrangements, for example, with Spain, Portugal, and Italy. A dislike by Irish clerics[1] for such an explicit and direct church–state relationship has a long tradition and comes out best in De Tocqueville's (1957) conversations with Irish clergy on his visit to Ireland in 1834. When asked if they would like subventions from the state to aid their stipends and church buildings, a move which was being seriously considered by the British government at the time, priests and bishops were united in rejecting the idea on the grounds that it would drive a wedge between clergy and people, identifying clergy with the principal enemies of the people. De Tocqueville's notes reveal not only the conscious opposition to such a mode of religious power but also how deep the solidarity between clergy and people was, the degree to which the poor, half the catholic population at the time, looked to the clergy for material and spiritual leadership, guidance, and assistance, and how much they trusted them.

However, with the emergence of the Southern Irish state, it soon became clear that this secularization of the state form was not to signify an absence of Roman catholic power in the construction of public morality, but rather an indirect recognition of the sovereignty of the church in most areas of moral concern besides education. In fact, precisely because Roman catholic power was to be accepted as normative in an entirely natural way by the catholic–nationalist population, the church's part-active and part-passive

[1] Such a dislike was no longer absent by the 1930s. When making proposals for the new Irish constitution discussed below, a number of catholic clerics tried to convince de Valera of the need for a Concordat between the Irish state and the Roman catholic church, but he refused to countenance it.

acceptance of the capitalist system of government and thorough opposition to socialism had a significant legitimating function for the Irish state-form. But this will not be the main focus of attention here; to make the general point requires tracing only the impact of the specific church teaching of the period on constitutional law and its implementation in Ireland. A Roman catholic ethos is not only present in the constitution of 1937 but has penetrated into affairs of state, legislation, and decisions over the destinies of individuals with frequency. A significant aspect of the religious hegemonic role can be seen in the translation of religious preoccupations into law via the concept of natural law. This mediation of ethics makes indirect one's access to the problem of the relationship between the religious faith of some and the morality of the many, particularly on the issue of the freedom of the individual versus the intervention of the state. In this respect, it is useful to pay attention to what we might now term the human rights' issue as one examines the handling of public–private morality in the Irish constitution.

The first constitution of the Southern state was the Constitution of the Irish Free State of 1922. Under the treaty with the United Kingdom, the twenty-six counties which were to become the independent state were to remain within the British Empire with dominion status, and the British monarch was to remain head of state. These principles were embodied in the constitution. The prime reason for bringing in a new constitution in 1937 was the return to dominance of the republican grouping, the side which had lost the civil war, and which had retained the intention of establishing a republic. The new Fianna Fáil Party, under the leadership of Éamon de Valera, was intent on asserting Irish independence to the full by leaving the empire and providing a president to replace the monarchy, thus becoming the Republic of Ireland. The 1922 constitution approved by Britain was considered a model of libertarian democracy. The preamble of the new constitution of 1937 proclaims:

In the Name of the Most Holy Trinity . . .
We, the people of Eire,
Humbly acknowledging all our obligations to our Divine Lord, Jesus Christ, Who sustained our fathers through centuries of trial,
Gratefully remembering their heroic struggle to regain the rightful independence of our Nation . . .
Do hereby adopt, enact, and give to ourselves this Constitution.

Clearly, the spirit of the laws of this Irish state was to be a religious one and therefore one which would not take account of the then one thousand, and now ten thousand or more of the population in the Southern state who professed no religion. But it was also an explicitly Christian spirit—unfortunate therefore for the four thousand Jews and for the tiny but now increasing numbers of Hindus, Buddhists, Muslims, and Taoists. The preamble was reinforced by Article 44, paragraph 1:

The State acknowledges that the homage of public worship is due to Almighty God. It shall hold His Name in reverence and shall respect and honour religion.

The presidential oath (Art. 12) also embodied this principle by calling on God's name explicitly.

Church teaching was already having an impact. The spirit of papal statements throughout the nineteenth and early twentieth centuries was that it was the duty of the state to oppose freedom of conscience in matters of religion and freedom of worship[2] and to celebrate openly the worship of God 'in that way which he has shown to be his will', namely Roman catholicism (Leo XIII 1903: 111–12). In a sense, the new Irish constitution was going against the spirit of this approach. As already mentioned, the Irish bishops had for some time found the separation of church and state both a workable and desirable solution. In addition, paragraph 3 of the same article recognized the main protestant churches then existing in Ireland. However, this was a fairly necessary compromise: paragraph 2 had already conferred on the Roman catholic church a 'special position' in the state, on the grounds that the church was 'the guardian of the Faith professed by the great majority of its citizens'. The concept of the majority is an important one in both British and Irish politics as already seen in Chapter 4. What matters for the moment is to examine the contention that the Roman catholic church had this special position only because of this matter of fact. It is true that the constitution was explicit in recognizing important personal rights in Article 40: equality before the law, the non-recognition of nobility privileges, defence of person and property, liberty, no detention without trial, freedom of speech and assembly. In fact, some of the paragraphs reproduced in summary

[2] See Gregory XVI in the encyclical *Mirari Vos*, 1882; Pius IX, *Syllabus of Errors*, Nos. 15, 78; and cf. Freemantle 1956.

form the personal rights articles of the constitution of 1922 and thus could claim to maintain the latter's libertarian heritage. Yet the articles which followed qualified certain of these rights in a typically catholic way and in the terms of the papal teaching of the day. Consider Article 41:

The State recognises the Family as the natural primary and fundamental unit group of Society and as a moral institution possessing inalienable and imprescriptible rights, antecedent and superior to all positive law . . .

This has a clear relationship with Pius XI's teaching, eight years earlier on the same subject: 'the very fountainhead from which the State draws its life, namely, wedlock and the family' (1929: 14), and with the dispositions of the then current *Code of Canon Law* which had come into effect in 1917: 'The marriage of baptized persons is governed not only by divine law but also by church law. The civil power only has competence in matters regarding the effects of such marriages' (Code of Canon Law 1917: 1016). The dispositions of the Irish constitution were clearly designed to accommodate such a position in full. The implications were spelt out in Article 41, paragraphs 2 and 3, which pledged the state to protect marriage and the family and forbade divorce. Thus were re-echoed the papal teachings of the indissolubility of both Christian and non-Christian marriage and of the state's obligation to reinforce that indissolubility (Leo XIII 1903: 68, 77, 78—*Arcanum Divinae* 1880; Pius XI 1930: 4, 16; Code of Canon Law 1917: 1013).

Article 42 recognized that 'the primary and natural educator of the child is the family'. Again, this view is related to Pius XI's, as expressed in *Divini Illius Magistri*: 'The family holds directly from the Creator the mission and hence the rights to educate the offspring, a right inalienable because inseparably joined to the strict obligation, a right anterior to any right whatever of civil society and of the State . . .' (1929: 14). Article 42 went on to give an almost supplementary right to the state in providing for a child's education by acknowledging the right of parents to school their children in their own home should they wish it. The article as a whole is strangely lopsided but seems to follow on from the logic of this position, embodying the agreement made between clergy and politicians towards the end of the nineteenth century outlined in Chapter 3.

The article now left the way open for the full expression of the teaching of Pius XI that, though the family had 'priority of nature and therefore of rights over civil society', education belonged 'pre-eminently to the Church, by reason of a double title in the supernatural order, conferred exclusively upon her by God himself; absolutely superior therefore to any other title in the natural order' (1929: 5–6). The double title consisted of the church's mission to teach under infallible guidance and her 'supernatural motherhood' by which the church 'educates souls in the divine life of grace' (1929: 7). As the church owned and managed the majority of the schools its faithful attended, and had a role in the management of the bulk of the remainder, the Roman catholic laity were left mainly on the receiving end of the educational policy thus legitimated.

More was to follow. There had been no mention of the individual's property rights in the 1922 constitution. In the new one, as if to repeat the church's social teaching against communism that 'every man has by nature the right to possess property as his own' (Leo XIII 1903: 210), Article 43 announced:

The State acknowledges that man, in virtue of his rational being, has the natural right, antecedent to positive law, to the private ownership of external goods.

Not only did the church's social teaching directly enter into the definition of rights, wrongs, and obligations within this nation state, but it was also to affect one aspect of the structure of government. The senate was to have 60 members: 11 were to be directly appointed by the Prime Minister, or Taoiseach, and 49 by election. Six of the latter were to be elected by the two Irish universities. But the remaining 43 were to be elected by local and national politicians on a vocational basis: that is, they had to be elected to five panels for which they would qualify by having the requisite vocational expertise—administrative, cultural–educational, labour, industrial and commercial, and agricultural. A somewhat unusual system one might think: but not according to the papal encyclical *Quadragesimo Anno* of 1931. It sought to overcome conflict within capitalist societies by encouraging Christian leaders to reconsider Leo XIII's suggestion of introducing 'industrial associations'. These were to be intermediary between capital and labour and based precisely upon types of vocational expertise in a way similar to medieval guilds (Leo XIII 1903: 240–7; Pius XI 1931: 32–4).

Reasons for the Political Religious Synthesis: First Conclusions

The way these various aspects of current Roman catholic social teaching were presented in the Irish constitution can hardly be taken as fulfilment of the need to reflect the opinions of the majority of the republic's population. The confessional and moral articles were largely the work of the constitution's architect, the then prime minister Éamon de Valera. He certainly felt that the culture of the state should reflect the fact that it was 93 per cent Roman catholic. But it is hardly possible to reconcile this with the claim of the constitution in Article 2 that the new state was the legitimate means of government for the entire island, which included over one million dissenting protestants, forming some 20 per cent of the total population. Their culture was not to be considered at all in the constitution. One is faced here with the blind spot of the dominant form of Irish nationalism, already so apparent in the preamble to the constitution itself, a blindness made possible by the ideological differentiation of state and religion combined with the ideological unity of the people, seen at once as both nation and catholic. De Valera was continuing the now dominant culture recognized by Eoin MacNeill and Douglas Hyde when, in his St Patrick's Day address to the United States of America, broadcast in 1935, he underlined the spirit of that preamble: 'Since the coming of St. Patrick, fifteen hundred years ago, Ireland has been a Christian and a Catholic nation. All the ruthless attempts to force her from this allegiance have not shaken her faith. She remains a Catholic nation' (*Catholic Bulletin*, 25/4 (Apr. 1935), p. 273, quoted Whyte 1980: 48). To the link catholicism-nation de Valera was also adding Gaelic identity in the form of the state promotion of the language:

The Irish language as the national language is the first official language.

The English language is recognised as a second official language. (Constitution of Ireland 1937: Art. 8, paras. 1, 2)

De Valera, was, however, openly aware of the alienation of Ulster protestants which such an affirmation of language implied:

I would not to-morrow, for the sake of a united Ireland, give up the policy of trying to make this a really Irish Ireland—not by any means. If I were told to-morrow, 'You can have a united Ireland if you give up your idea of restoring the national language to be the spoken language of the majority of the people', I would, for myself, say no. (Quoted Laffan 1983: 119)

Whether or not he was equally aware of a similar alienation process in asserting the catholic nature of the nation is perhaps more open to debate. In any case, the practical attitude clearly excluded the protestant population of the island and left protestant loyalists in no ambiguity as to their place in any future united Ireland. If Ireland was a catholic nation, then its law and government would be correspondingly so whatever the disposition of church–state relationships, theocratic or secular.

The second likely reason for the deliberate use of church teaching in formulating parts of the constitution is implicit in the presence of the papal-oriented articles of the constitution themselves. Pius XI had made it clear in 1930, subsequent to the success of the Vatican-Italian Concordat of 1929, that good sons of the church with political power were to look to the church itself for guidance in their statesmanship:

Governments can assist the Church greatly in the execution of its important office if, in laying down their ordinances, they take account of what is prescribed by divine and ecclesiastical law, and if penalties are fixed for offenders . . . As Leo XIII has already so clearly set forth: '. . . The dignity of the State will be enhanced, and with religion as its guide, there will never be a rule that is not just . . .' (1930: 64)

In fact, de Valera and the civil servants he appointed to assist in the work, consulted Irish theologians on matters of society and church–state relations (Longford and O'Neill 1970: 295–6; Whyte 1980: 379; Keogh 1987; Faughnan 1988). The most important was John Charles McQuaid, a Holy Ghost priest and later archbishop of Dublin. Jesuits and other Holy Ghost priests were also involved. Papal encyclicals came to be consulted by this route, as did the current treatises of *Ius Publicum* on church–state relations. These were heavily biased towards the view that there were two perfect societies, both supreme in their own orders, with the church supreme in spiritual and moral matters (see Pius IX's *Syllabus of Errors* in Freemantle 1956: 145–6, 152; Leo XIII 1903: 114; Pius XI 1930: 65; for typical treatise approaches see Van Noort 1960: 235–41; Bender 1960).[3] Of course, it was for the church

---

[3] Keogh (1987: 21) sees McQuaid's proposals in this area heavily influenced by the triumphalist Fr. Denis Fahey, who had been his teacher. However, a good deal that McQuaid submitted is typical of the *Ius Publicum* and papal encyclicals of the time, as much reflecting Vatican attitudes as those of the French triumphalists on whom Fahey relied.

to decide what constituted a spiritual or moral matter. But, to his credit, de Valera toned down the catholicism in the drafts suggested to him by the clergy, consulted Rome, and was successful in getting it at least to be neutral about his preferred formulations (Keogh 1987: 43–53). He also consulted protestant church leaders on the religious elements of the new constitution, a step unprecedented in catholic Ireland of his day, and gained their support on the basis of the recognition of their churches the constitution was to give.

But there is a third and underlying reason which must have motivated de Valera and the religious and political intellectuals in this religious pursuit of an allegedly secular and public morality which mirrored Roman catholic teaching. The issue of heeding the church was not simply one of seeking a mediator of divine guidance on matters of Christian conscience. For the Roman catholic social teaching of the day laid down certain things as of natural law which anyone who was reasonable and honest was supposed to be able to recognize. The advice of the church's moral experts and authoritative clerics was thus pertinent to the entire population of the world, let alone that of Ireland. The conscientious lay person was simply involved in a two-way process of having his or her conscience informed, while at the same time being protected from making unintended errors of judgement. The church was acting on behalf of all, and had a prime obligation to push for a public morality which matched its perception of the natural law lest the very fabric of society be torn asunder and its members cast on the road of moral decline. What was good for catholics was also good for protestants, and Ireland was to be no exception.

In fairness to de Valera, it must be said that he opposed any attempt to incorporate the church into the apparatus of the state and in this was, as already noted, going against the form of relationship preferred by the Popes of the day. But his attitude was by now part of the common sense of clergy and people alike. There appeared to be no reason to upset long-established routines in this respect. But that did not mean that the ethos of a Roman catholic state would be in any way diminished, only that the form of political religious power would be different.

An important aspect of the successful implementation of these parts of the constitution was that the political opposition did not object to them. They really were part of the unquestioned reality of the culture of the state. Part of this culture was a certain distinction between what made up the sacred and what made up the profane spheres. If guidance were to be sought in matters which hovered on the boundary between the two, it was likely that the church would be seen by people and politicians as authoritative in resolving the issue. But there was in any case no doubt that the areas of the family, education, and public morality constituted part of the sacred sphere.

One explicit guiding principle of papal teaching which came to influence the Irish constitution and successive debates culminating in the church–state imbroglio of 1951, dealt with below, was that of 'subsidiarity': that no 'secondary' institution such as the state should take on duties which 'primary' and 'intermediate' ones, such as the family, could assume. Secondary institutions were also expected to assist primary ones in fulfilling their essential tasks. The principle took on particular importance in the light of the development of totalitarian communism, and the Roman church saw the threat to be endemic in the growth of the modern state apparatus, including its welfare institutions. It is also important to notice how easily such a principle supports personal enterprise and property in its contemporary form, capitalism.

Aspects of the Irish constitution and its implementation are clearly oppressive as well as offensive to other minorities besides the protestant ones. The structure of this oppressive power contained not the Roman catholic church as state executive, but as what Gramsci termed 'an organic intellectual', as a planner and co-executor of the ethos of the state, merging this public aspect of the catholicism of the day with ideological elements from more recognized sources of catholic-Irish nationalism: the language movement, anti-imperialism and, generally, catholic-Irish domination.

The purpose of detailing the relationship between the Irish constitution and the papal teaching of the period has not been to ridicule seemingly outdated church–state dispositions. The writings of Leo XIII (1878–1903), on which those of Pius XI (1922–39) were frequently constructed, are remarkable expressions of a papacy dragging itself up by its bootstraps from centuries of neglect in the field of social, economic, and political issues. It had continually opposed socio-political change and had had little understanding of the industrialization process up to that time. The growth of the catholic social movements and activity of Roman catholics in trade unions and politics was rooted in this dramatic revival, which gave rise to Catholic Action and the development of catholic social studies groups around the world. The birth of *sociologie religieuse* and the initial contributions of catholic institutions both in Europe and the USA to the founding of departments of sociology are rooted here rather than in the work of the founding fathers of the discipline, though clearly there are links between de Bonald, Maistre, and Durkheim on the nature of social solidarity and the concept of society, and authoritative Roman catholic social theorists.

Equally, the stress on the negative influence of Irish catholicism is not meant to imply a negation of its positive aspects: its sustenance of the people during long years of oppression, the dedication of its clergy and religious orders, both in education and in welfare, not to mention the contribution of its missionaries to the fight for social justice in third-world countries. Its lay politicians have equally made important contributions. De Valera himself resisted right-wing pressures to enhance the role of the church in the Irish state and successfully opposed tendencies to fascism apparent in Irish paramilitary movements of the thirties (Manning 1970).

The Pre-Constitutional Ethos: The Case of the Protestant Librarian of County Mayo and Issues of Sexual Morality

Was the Irish constitution merely an abstract one, without actual social consequences? It was not only a constitution which really was and is effective in the running of the state, but which was itself the product of the hegemonic culture already established. Whyte (1980) provides several instances of the influence of the Roman catholic moral code in both state decision-making and in public behaviour from 1922 through to the 1950s. The most interesting one, in which clergy, local people, and the Dáil all played their part, was the case of the protestant librarian in County Mayo, 1930–1. A woman protestant was appointed as county librarian, but when the county library committee met to ratify the appointment, it refused to do so by ten votes to two. The committee was largely made up of Roman catholic clergy, and they clearly opposed the appointment on the grounds that a protestant could not be trusted to safeguard the catholic morals of her charges in her purchase and loaning of books. Though the Cosgrave government then in power insisted on the appointment, the local population and authorities boycotted the operation of the library, and the government had to concede by moving the librarian elsewhere. When the matter was debated in the Dáil, Fianna Fáil opposed the government view, and de Valera himself supported the local population's opposition on the same religious grounds.

From the point of view of the issue of law, there can be no doubt as to the influence both of the constitution of 1937 in the period following its enactment and of the influence of Roman catholic teaching on legislation prior to that date. But these were all in areas circumscribed by the church as spiritual matters. The church's greatest denunciations were reserved in the early period for sexual immorality and the dangers to the people coming from modern society: from the cinema, literature, radio, and, above all, from dance-halls (Whyte 1980: 25–34). Already in the eighteenth and nineteenth centuries, there had been a powerful alignment between rural cultural ethics and church teaching. On the sons and daughters of the farming family was imposed the strict necessity of abstaining from sexual activity other than for the first born, so as not to impair the integrity of land succession. From the church came a dual impulse strengthening this morality: the prevailing teaching on sexual abstention outside of marriage under pain of mortal sin, and the rigorous life of a clergy pledged to chastity and preaching the need for an unmarried laity to practise the same degree and kind of circumspection in sexual matters which the clergy had been taught to impose upon themselves. If anything, the return of the land to the practising farmers by the land Acts would have strengthened such a morality.[4] It is also interesting to note that the issue of land inheritance was a key reason why the Irish population voted 'no' in the referendum on divorce in 1986. That the local priests continued to play a significant part in the regulation of local sexual morality from the beginning of the state is shown by accounts of 'boy meets girl' from the 1920s. For example:

In 1931 we got a new P[arish] P[riest]. He condemned dancing in every form, even the kitchen dances were sinful and against the wishes of our Church. Boys and girls should not be on the road after dark. The C[atholic] C[urate] was sent out to patrol the roads and anybody found or seen on the roads had to give their names. The people who allowed boys and girls into

---

[4] For some aspects of the debate on church influence on sexual mores in Ireland see Connell 1968.

their home to dance were committing a grave mortal sin . . . Dancing was the devil's work . . . (Kavanagh 1954, quoted Whyte 1980: 29–30)[5]

The Irish state apparatus assisted public and family morality by the fairly heavy censorship of books, magazines, newspapers, and films.[6] It also banned the making, importation, or distribution of contraceptives, and the distribution of any literature informing on such matters. The reaction of the first Dáil to successive attempts to introduce legislation for divorce was to change standing orders so that such a bill could not be introduced again. Since the new constitution was enacted in 1937, the prohibition on divorce has become so strict that couples who have been granted a nullity decree by Roman catholic canonical tribunals have found that they may still not be recognized as single by the state, and are thus unable to remarry in the Republic. There has been a degree of attenuation of this rigidity in some respects, but only in favour of the church's interpretation. People seeking to remarry must first get a divorce abroad—which in practice means Britain. Naturally, apart from the accusation of hypocrisy and of letting Britain solve its marriage problems, the government ends up being accused of allowing divorce for the wealthy and none for the poor, as the cost of going to and staying in Britain and paying for the legal procedures is beyond the pockets of the majority.

An additional concern of the church remained the fear of socialism. The labour party, which still had some of the socialist leanings of its founder James Connolly, was forced by its own lay members to amend its draft constitution in 1940 in order to fall in line with current papal social teaching. Phrases such as striving for the setting up of a 'Workers' republic' and the establishment of 'public ownership' were excised, as they were seen by the members of an affiliated teachers' union, and ultimately by the bishops, to counter church doctrine (Whyte 1980: 82–4; Keogh 1982: 7, 77 n. 5). The teachers' union even submitted the draft constitution to the bishops so as to get it right. It is to be noted that, in this, it was the laity who took the initiative.

---

[5] Kavanagh 1954, as reported by Whyte 1980: 29–30. It is interesting to note that by the 1960s in Italy, it was only dancing in public dance-halls and out of the view of parents or overseers which still remained mortal sin—this teaching formed an essential part of the training of seminarians for the confessional in Italy, a process the writer himself experienced.

[6] Censorship of Films Act, 1923; Censorship of Publications Act, 1929.

Testing the Alliance 1: The Adoption Controversy

The way in which the morality received by the state from the Roman catholic church could create conflict within the state but still be transferred into the coercive sphere of state activity can be seen in two cases in particular. They are indicative not only of the strength of current Roman catholic social rules and sense of obedience on specified issues at the level of popular religion, but also of the direct power of the hierarchy operating through the state in affairs they considered sacred. The two cases also show that change is a possibility, but that it has mainly been in the past through change in the hierarchy's and clergy's attitudes as well. The cases equally indicate the boundary beyond which the church was seen as abusing its definitions of the sacred.

The first of these incidents occurred with the attempt to introduce a law on adoption. During the period 1947–51, an inordinate delay on the part of the government was experienced in bringing forward the requisite bill legalizing adoption, despite considerable pressure from prospective adoptive parents and adoption agencies. In the new state, fostering was the only way to provide a home for orphaned and otherwise parentless children. It became clear that the relevant government ministers were deliberately hindering the progress of the necessary bill. When both the Minister for Justice, General McEoin, and the Attorney General, Mr Charles Casey, made it clear that they considered the legislation inopportune, pressure for an explanation eventually brought forward a response from Mr Casey. He argued that Article 42 of the constitution claimed the primary right of care for the natural parents and was inalienable: it could not be delegated.[7] In addition, he argued that legalizing adoption was against the teaching of 'the church'. He further argued, or at least implied, that to permit a 'non-catholic' to adopt children would be wrong because it might put such children's eternal salvation in jeopardy (Whyte 1980: 188–9). It later became clear that the two ministers had consulted the Roman catholic archbishop of Dublin before delivering their judgement on the matter.

An adoption bill eventually became law two years later, in 1952, under the next administration. Mr Casey and the archbishop were

---

[7] This interpretation of the constitution was only overruled by the Supreme Court 15 years later, in 1966.

judged to have followed a conservative catholic viewpoint and the 1952 solution was adopted because it reflected what had become a more theologically certain position on the matter. But even the new bill limited adoption to parents possessing the same religious identity as the child. This meant that couples who had married across the religious divide were not allowed to adopt at all. As will be seen in Chapter 7, this itself could only have been due to current Roman catholic social teaching on mixed marriage. It was 1974 before that particular form of discrimination was removed by a new Adoption Act.

Testing the Alliance 2: The Mother and Child Scheme Controversy

The second case reveals even more the extent to which the legitimation of the state was a Roman catholic affair, at least as one of the two principal sources of power in the alliance.[8] From the early 1940s, the Irish government began to work towards the introduction of a comprehensive health service for mothers and children in line with other legislative developments in Western countries. There can be no doubt that the lack of such a programme bore heavily upon the poor, and that poor health and mortalities were a consequence. In 1946 there were already signs of clerical opposition to any socialization of welfare in queries about Fianna Fáil's proposals from Archbishop McQuaid and the Confederation of Convent Schools. As has already been seen, it was the style of both church and politicians to avoid their mutual consultations being known, which tells us that the secularity of the state at that time was partially a façade, but one which it was felt by both interested parties had to be maintained, probably so as not to confuse the faithful. In a letter to the Irish premier in 1947, while an extensive health bill was going through parliament, the bishops 'pointed out that to claim such powers for the public authority, without qualification, is entirely and directly contrary to Catholic teaching on the rights of the family, the Church in education, the rights of the medical profession and voluntary institutions' (*Irish Independent*, 12 Apr. 1951, quoted Whyte 1980: 143). It appeared that one of the main sources of the bishops' opposition was the authoritarian nature of the legislation. That the bishops were

[8] The classical historical account of the controversy is Whyte 1980: 196–302, who also provides a section of relevant documents in his App. B (419–48). See also Barrington 1987: 167–221; McKee 1986.

affected by the opposition of the medical profession to the scheme—clearly they stood to lose money and independence if such a scheme were implemented—seems likely. Whereas similar provisions in other countries were mainly designed to provide facilities which could or could not be used, the Irish ones were to be compulsory and there was to be no choice of doctor. The hierarchy feared the move as the thin end of the wedge of secularism and the totalitarian state. It appears they feared the appointment of secular medical personnel who might educate mothers in family planning and their children in sexual matters. Additionally, they feared the full development of a national health service, viewed by them as unacceptable socialism. A change of government in 1948 found the new administration equally committed to health reform. The new Minister for Health, Dr Noel Browne, a dedicated reformer of the health services and much concerned in particular with the eradication of tuberculosis in Ireland, modified the earlier bill to exclude the compulsion elements. But by March 1951, it was clear that even this reform did not satisfy the Irish hierarchy who had also been lobbied by the Irish Medical Association to oppose the scheme. Grounds for the bishops' opposition were that only parents and not the state should have the right to provide for the health of their children, that the state had no role to play in the physical education of children and mothers, and that individual privacy would be threatened by public use of their private health records (Whyte 1980: 213–14).[9] Later the hierarchy made it clear that the basis for their objections was the increase in power by the state *vis-à-vis* the liberty and self-reliance of its citizens which such legislation would entail.

Perhaps the most revealing aspect of the whole affair is the way Dr Browne handled it. The members of his party appeared to know nothing about the opposition of the hierarchy to the earlier Bill, and Dr Browne only found out in October/November 1950. When he was told of the hierarchy's opposition to any bill of this kind, Dr Browne, himself secularly inclined, submitted his proposals and arguments to the hierarchy for approval, just as a previous Health Minister, Dr Walsh, had consulted Archbishop McQuaid on a similar matter in 1946. If a man of such convictions felt it necessary to follow this course, how much more would have his religiously

---

[9] Most of these objections had already been stated in the bishops' letter of 1947 to the Irish premier. See Barrington 1987: 186.

devoted colleagues. In fact, his cabinet colleagues, already unsure about public support for the bill and alienated from Dr Browne by his provocative politics, were not prepared to go against the hierarchy's condemnation and refused to support the proposal. They moved that a new scheme should be prepared, in conformity with catholic social teaching. Dr Browne was forced to resign.

Dr Browne has been something of a maverick in Irish politics even up to this present time. He took the unprecedented step of going public on the issue, an action of which Whyte does not sufficiently stress the importance. He handed on the correspondence between himself and the Bishops to the *Irish Times*, the recognized liberal and 'protestant' newspaper of the republic. It was a contentious gesture. Many were shocked and surprised at the interference of the bishops in the parliamentary process. Others resented Dr Browne's tactics, for the church was clearly not to be brought into such public disrepute. Again, it was clear that the ministers and their parliamentary colleagues, many of whom in any case disliked the scheme, would not pass into legislation any bill which had been declared contrary to church teaching by the bishops. At this level, the power of the Roman catholic church bore directly on the institutions of government. Dáil members would act on a decision of the hierarchy once that hierarchy had said the teaching of the church was at issue, providing of course that the area in question was a grey one, a new territory not previously covered by the generally accepted sacred-profane spheres. In such an unclear area, it was not to be left to Dáil members to decide, irrespective of whether it still remained an appropriate public measure. Even if Dáil members had thought otherwise, it must by now be clear that the ethos of the Irish Republic was still one in which it was impolitic to be in conflict with the church.

Reasons for the Political Religious Synthesis: Second Reflections

When the next government under de Valera introduced a very limited scheme of public health assistance in 1952, the hierarchy still opposed its introduction, and by direct methods as well. Only de Valera's personal intervention convinced them of their lack of perception and judgement. It was the 1960s before the bishops began to distance themselves from direct relationships with the government of the day. It could be argued that such a strategy was in any case unnecessary. However, it was not simply a strategy, but embodied a particular ideology: the belief that the bishops were the church above all, and that they, not the laity, were the ones to communicate with the state. The desire to make arrangements covertly could well have been affected by a double pressure. On the one hand, the laity would not understand the complexity of the issues because of their poor understanding of Christian doctrine and the intricacies of morality; on the other, the area of the sacred in the political arena was clearly defined, and clear guidance from the clerical church was both necessary and expected by good Christian politicians. What the bishops and the politicians had come up against in the Mother and Child controversy was that this paternalistic conceptualization was intrinsically at odds with the common understanding of democracy. This appears to be the first contradiction which the bishops later sought to resolve by withdrawing from the direct contact method. It indicates that a real change did eventually take place after all, in so far as freedom of information and the inappropriateness of direct consultation without public mediation eventually became accepted as something which, at worst, had to be risked and something which, at best, formed an essential task of the construction of a Christian conscience in a Christian society.

This, however, was the only contradiction which appears to have been solved by the 1960s. The popular disquiet had deeper roots. One was the detestation by the liberally oriented of religious paternalism, a mild form of anti-clericalism. The other was the sense of injustice experienced by the subordinate classes at seeing the scheme dropped. Many of these had just as much sense of loyalty and devotion to the church as the middle classes. Large sections of the population, the subordinate classes specifically, were desperately in need of improved health care. Many of them appear to have been angered and disappointed at the church's suppression of the proposed scheme. The fact that it was not long after that similar if lesser welfare legislation was introduced is indicative of the public demand and that clerical socio-moral theory did not tally with the people's experience of reality.

The case reveals the extent to which the church as an institution was coupled with the nation. Part of the state's legitimate authority rested on this acknowledgement and it was built into its hegemonic structure and consensus. The case highlights the extent to which it was possible for the religious intellectuals to misjudge exactly where the consensus lay, as well as the extent to which an evolving historical situation which created areas for new decision-making provided the field for a new power contest. It was then more than a misjudgement: it revealed the extent to which the religious intellectuals' theory of church and state was outdated and incoherent.

What might have been the nature of the inner religious conflict? One could try to approach it with the notion of spheres of competence. The actual sphere of competence the church had sought to monopolize had been that of sexual morals, temperance, the semi-private sphere of the family, and education. It was when the semi-public nature of family matters became the subject of attention that the sources of the legitimacy of the state—the Irish nation, the church, democracy—were brought into conflict. To be more precise, the conflict was between the felt need of interventionism by the majority of the population and its politicians and the particular social theology of the churchmen. The churchmen had failed to provide another theology suited to the case. They failed to see that the principle of subsidiarity as interpreted was inadequate to a modern society. The process of industrialization and the accelerated growth of the division of labour had already wrested from the family, including its rural variant, its monopolistic role in the processes of production. The family had changed, and its various functions of education, food production, and the manufacture of clothing were already in part relinquished to a variety of institutions whose smooth functioning rested on the intervention and guidance of the nation state. The change in the socio-moral context was hardly recognized, and the corresponding change in religious social ethics had remained unformulated. In the case of the Mother and Child scheme, it was rather through withdrawal from the scene that first the politicians and later the churchmen avoided further trouble.

As will be elaborated now in further contexts, it appears that the underlying reality to such difficulties was, and remains a complex contradiction which the catholic–nationalist alliance has still failed to solve. On the one hand there were the beliefs in the nature and extent of the clergy's political religious power and how that power was to be exercised in the state. There was also the general assumption that the fact of a state populated by catholics must perforce imply a heightened degree of institutionalized catholic value in law. On the other hand, there was the nationalist theory of democracy, which derives power from the people and not from an élite group of religious intellectuals. The next section will show the way in which this contradiction continued to be shared by a large number of clergy and laity even into the 1980s. It should be added here that the high clergy hardly recognized that they had actually been exercising a political religious power of a specifically sectarian or monopolistic type. Hence when Irish clergymen stated that church and state were separate, they meant they themselves had no legislative or executive role in the state. But they did and still do remain the authoritative conscience of the nation.[10] In addition, the clergy appear to have considerable difficulty in recognizing the transformation which theoretical positions on Christian belief and morality undergo as they are concretized in historical human relationships, doubtless also because of the strong essentialist bias in their perception of socio-ethical issues.

Part of the overall argument of this book is that, as the Roman catholic church is principal validator or legitimator of the Southern state along with the concept of the national entity, what that state goes on to do in the field of social ethics cannot be separated out from the responsibilities of the church. This structural link between politics and religion is underlined by the legitimacy the church gives to the state as a whole, by recognizing it simply as society. However, it is the further legitimation given to a particular socio-ethical form of political religion in the constitution and to the implied authority of religious intellectuals in deciding *in concreto* what must be taken by the state as in the interests of the common good which gives to conflict in Ireland between the two alliances its political religious dimension in the sphere of law. It is almost as if the catholic Irish people become a people of God in Old Testament terms, in a way similar to the Northern Calvinists, a people which have overriding power to set up their domain.

The continuing permeation of civil society by this particular variant of the Roman catholic ethos was most natural. As already stated, the clergy had sustained the catholic–nationalist populace over centuries of oppression. Furthermore, the ranks of that clergy were drawn from the common people themselves. But a key element remained the considerable filial loyalty the catholic nationalists showed towards their clergy, bishops, and Popes. The

[10] Ryan (1979) appears to have been the first person to use this term.

authority of the clergy had been second to none during the years of persecution and this appears to have strengthened as a direct result of Roman centralization from the 1850s onwards. This respect for authority has appeared throughout the examples given, but it has also been shown to be a respect within certain limits.

RELIGION AND LAW IN THE SOUTHERN STATE, 1973–1986

Minor Indications of Political Religious Changes

There is a strong case for arguing that the dominance of Roman catholic high clergy in the determination of key value ingredients in catholic–nationalist consensus in the South is in a state of change, and that certain aspects in particular are being challenged. From the 1950s onwards there was considerable improvement to the standard of living. As Ireland opened up to foreign investment under de Valera's successor, Sean Lemass, another element in the value structure came to prominence, namely the obligation to achieve and retain full employment within the context of a modernized economy and a satisfactory level of housing, welfare, and medical care for all. The economic growth of the 1960s and early 1970s came to grief because of the western economic slump of the last decade and the massive increase in the Southern population. This grew from three to three and a half million by 1981 and was created by a combination of net immigration, earlier and more frequent marriage, and the maintenance of relatively high levels of childbirth. This is a major reason, in addition to state interventionism, why economic matters precede the issue of Northern Ireland both theoretically and practically in contemporary Irish politics.

A further change has been the growth of liberal theology within the Irish Roman catholic intelligentsia. The influence of the second Vatican Council (1962–5) has enabled theological discourse to become dominated by non-hierarchical models of the church and a generally eirenic theology in which all the churches in Ireland share. Roman catholic theologians teach at Trinity College, former bastion of protestant Ireland, and Roman catholic writers, including clergy, question the agreed wisdom of the Irish catholic hierarchy. But, as will be seen, such liberal theologians have yet to achieve power within the political religious establishment.

Criticism of the Irish constitution itself from both clergy and laity produced one first change in the constitution. The famous paragraph of Article 42 intending to give pride of place to the Roman catholic church in the state was removed by referendum in 1972, along with the accompanying paragraph recognizing other existing religious groups. The Roman catholic hierarchy declared that they had no opposition to this particular change, but one wonders how the vote would have gone if they had. Also, in 1974, an adoption Act made it possible for couples of mixed denominational or religious origin to adopt. But Roman catholic adoption agencies, the source for the majority of children awaiting adoption, are not obliged to assign baptized Roman catholic children to such couples, and whether they do or not is not a matter of public knowledge. The episode shows, as Whyte says, 'not a clash between Church and State but a shifting consensus involving them both' (1980: 399).

A further significant change in political religious attitudes occurred with the liberalization of legislation prohibiting contraceptives to be bought or sold, or even imported into the state. A campaign begun in 1971 to repeal the ban dating from 1935 culminated in the famous appeal of Mrs McGee to the Supreme Court to declare she had a right to use contraceptives. The supreme court found in her favour in 1973 and declared the relevant section of the 1935 Criminal Law Amendment Act to be unconstitutional. However, a law dealing with the national situation as a whole was only introduced in 1979. The Act, while permitting the sale of contraceptives, restricted sellers to chemists and only on the production of a medical prescription. This did not prevent the Family Planning Association in its Dublin and Cork clinics from providing contraceptives freely and students at University College, Dublin from installing a condom-dispensing machine in their Union building. In fact, the need to liberalize the law was mainly because it was inoperable: thousands of people were already using contraceptives. A committee of the Irish Medical Association put the numbers of women using the contraceptive pill in 1978 as 48,000; and the Family Planning Association saw 30,000 people in 1976 and more than 53,000 in 1978 (Whyte 1980: 403–4).

It is quite possible that a number of other political changes would have got under way more recently were it not for the direct intervention of Pope John Paul II and the Vatican. In 1985, Rome declined to follow the convention of choosing the new archbishop of Dublin from the names submitted by the clergy and notables of the diocese, and imposed a known conservative and former president of Maynooth College, Dr Kevin MacNamara. No clearer way of indicating the path to be trod could have been made. It is possible that such authoritarian actions will occur again as soon as another crucial appointment to the hierarchy comes up.[11] Of course, Irish clergy and laity are sometimes at the forefront of political religious change in other countries, and a lively Irish intellectuals' religion will continue: but whether or not it will eventually affect the structure of power is another matter.

Major Indications of Conflict: The Abortion Debate

The shifting consensus of church and state was not the major feature of the period. The two changes which took place in law and constitution before 1980 were at least permitted by the clerical church in that it agreed with the abolition of the special position of the Roman catholic church, and did not oppose the limited introduction of contraceptives. They were both cases expressing the consensus of politician and priest. The major debates after that time have tended to show a split between church leadership and a large section of its laity on the one hand, and a significant proportion of Roman catholic laity, other churches, and secular groups on the other. In this confrontation, the traditional religious forces, which support the old monopoly religious structure, have so far prevailed.

The first major demonstration of this was in 1983, in the controversy over the abortion amendment. Two years previously, a campaign to write into the Irish constitution the existing legislation criminalizing abortion had got under way under the name of the Pro-Life Amendment Campaign. It is difficult to describe this body accurately. The bulk of its organizers were members of traditional organizations of the old sodality type: members from the Legion of Mary, the Knights of Columbanus, and one or two other organizations of more recent origin devoted to the strengthening of catholic morality. Having acquired a sufficient number of signatures, the Campaign submitted their referendum proposal, which the government of the day, the Fine Gael–Labour coalition under Garret FitzGerald which had come into power that year, decided

---

[11] In 1988, a new archbishop of Dublin, Desmond Connell, was appointed. The writer does not know which precise steps were taken, or if Rome interfered in a way similar to the MacNamara appointment.

to submit to the populace. The Roman catholic bishops supported the wording of the referendum and, as the referendum approached, said it was legitimate to oppose the amendment even if one was against abortion, which was in any case still immoral and was 'the direct taking of an innocent life' (Irish Bishops' Conference statement, quoted O'Carroll 1983).

However, on the eve of the referendum, the majority of priests preached against abortion. The nexus appears to have been made for a sufficient number of the laity, as those who eventually voted in the referendum voted in favour of inscribing an appropriate anti-abortion clause in the constitution. Article 40, paragraph 3, section 3, now reads:

The State acknowledges the right to life of the unborn and, with due regard to the equal right of life of the mother, guarantees in its laws to respect, and, as far as practicable, by its laws, defend and vindicate that right. (Ninth Amendment of the Constitution Act, 1984)

Several points of note should be made. Firstly, only slightly more than half the electorate voted, when other turn-outs for referendums have been around two-thirds; and despite considerable pressure from the clergy, only one-third of the total electorate actually voted for the measure. There is the suggestion, therefore, that the traditional popular religion is in decline, albeit still responsive enough to hierarchical religious leadership to produce the necessary political religious results. Secondly, anti-abortionism is the view of the great majority of the population in both parts of Ireland, as well as the various churches. The appropriate committee of the presbyterian church in Ireland opposed the extension of the British Abortion Act of 1967 to Northern Ireland (Presbyterian Church in Ireland 1930–86, *Annual Report* (1982), pp. 194–6). In consequence, it was resolved 'That the General Assembly is firmly opposed to indiscriminate abortion, but does not believe it is wise to insert a clause banning abortion into the Constitution of the Republic . . .' (Presbyterian Church in Ireland 1930–86, *Annual Reports* (1983), p. 9). Its position on abortion itself was hardly ambiguous: the minutes of the Assembly meeting for 1982 point out '[the Assembly's] opposition to abortion on demand for purely social reasons, or as a means of birth control . . . [and] that in exceptional circumstances, where medical abortion might be necessary, the most stringent safeguards should be provided to prevent abuse' (Presbyterian Church in Ireland 1930–86, General Assembly (1982), p. 61). The Church of Ireland maintains its adherence to the Lambeth Conference decision of 1958 to oppose abortion on anything but the most exceptional grounds (Church of Ireland Submission to the New Ireland Forum (1984), p. 2). The Methodist Church in Ireland was in session during an unsuccessful attempt by the government to quash the running of the referendum. It welcomed

a belated attempt to meet minority principles in a plural democracy by avoiding any definition and by restoring ultimate power of decision to the Oireachtas [Parliament]. The Conference regrets the failure of the attempt. It regrets that the previous Government [of Fianna Fáil, under Charles Haughey] made no concessions to Protestant appeals for a more open position . . . The Conference would wish all who may take part in the referendum to recognize that Protestant Churches are pro-Life but anti-amendment and to query whether they wish a clause in the constitution unacceptable to Protestant churches. (Methodist Church in Ireland 1983: 48)

The position taken by the protestant churches, and the newspaper articles and television debates which preceded the referendum appear to have affected public opinion. Caught between the preaching of the Roman catholic clergy and appeals from protestants and liberal Roman catholics—including Dr FitzGerald—not to constitutionalize the issue, it would appear that many voters simply stayed away rather than vote against the proposal. As mentioned, it was already forbidden by law to procure an abortion. Its constitutionalization is a direct result, not of this already widespread opinion, but of Roman catholic pressure of the monopoly type.

The writing into the constitution of the prohibition did not mark the end of the campaign for some groups. Legal action was taken in 1986 by the Society for the Protection of the Unborn Child (SPUC) to prevent the two family planning clinics in the Republic from counselling abortion. The high court decision, announced at the end of December in that year, found on behalf of SPUC and declared such counselling by the Dublin Well Woman Centre and the Open Line counselling service to be illegal and ordered them to cease their public information service on the issue by 12 January 1987. As the ruling only applies to the institutions named, SPUC intends to get further banning orders made against other groups and named individuals as and when it becomes 'necessary' (*Irish Times*, 30 Dec. 1986). It should be noted that contraceptives which can result in an abortion, often termed 'abortifacients' and including devices such as the coil and spermicidal lubricants and jellies, remain illegal in Ireland.

The Roman Catholic Hierarchy and the New Ireland Forum

The next significant event of constitutional relevance was the important initiative taken by the nationalist parties in Ireland in 1983–4, the New Ireland Forum. This was an initiative sponsored by the SDLP at first intended to bring together politicians from both catholic–nationalist and protestant–loyalist groups to discuss the future of the island as a whole. In reality only the major nationalist parties participated. Loyalist parties refused to join in, and the Provisional Sinn Fein was excluded by the fact that the ground rules for the forum only allowed parties accepting constitutional, non-violent politics to participate. The forum completed its consultations and reported in the Summer of 1984. There was one vital breakthrough for the nationalist consciousness, though there still remains some doubt as to how deep this was for the Fianna Fáil party: all the parties committed themselves to the recognition of the protestant–loyalist group as having legitimate aspirations and affirmed the need for the recognition of 'two sets of legitimate rights' coexisting in the island as a whole: 'Constitutional nationalists are determined to secure justice for all traditions ... The new Ireland must be a society within which, subject only to public order, all cultural, political and religious beliefs can be freely expressed and practised.' This was made explicit in terms of the sense of British identity which protestant loyalists experience: 'This implies in particular, in respect of Northern Protestants, that the civil and religious liberties that they uphold and enjoy will be fully protected and their sense of Britishness accommodated' (New Ireland Forum 1983–4: i. 22–3).

For the very first time, the leadership of all the major constitutional nationalist parties in Ireland, representing 90 per cent of the nationalist population, had abandoned the concept of the tyranny of the majority will. The people of Ireland were not identified as the catholic–nationalist population. There were still traces of ethnocentrism in the nationalist viewpoint elsewhere in the report: for instance, they still referred to 'the historic integrity of Ireland' (New Ireland Forum 1983–4: i. 28), implying an almost naturalistic concept of Irish unity, when as a political unit Ireland only ever existed as a British-administered territory. But whether the statement is rhetoric or reality, particularly for Mr Charles Haughey and the rest of the Fianna Fáil leadership, we will have to wait and see.

The Irish Roman catholic church leadership also submitted statements on its political religious position. While preparing an early draft of the present chapter, the writer heard Bishop Cathal Daly of Down and Connor, a leading moderate among the bishops, respected by all except PIRA, repeat the adage that the Irish constitution was secular, implying that there really was little to be done for the moment. This in fact was the main tack in the bishops' attitude to the forum. There appears to have been a general feeling among them that there was no real problem on this score. Concessions would only be required once a united Ireland was being negotiated (New Ireland Forum 1983–4: xii. 11).

Via the channels of public announcements at times of political decision-making, the bishops seek to enforce official Roman catholic morality on all members of the state. Their intention to continue with this strategy remains. Both in the Forum Report (New Ireland Forum 1983–4: xii) and in their written submission (Irish Episcopal Conference 1984), they not only repeated their oft-stated attitude to catholic schools, defending them from any contribution to sectarianism in Ireland, but opposed the introduction of divorce and any weakening of legislation which they felt protected the family. They explicitly quoted sections of Article 41 or the 1937 constitution so as to support the view of including its provisions in any new all-Ireland constitution. They further expressed reservations about the concept of pluralism and maintained an objectivist stance on the relationship between the Roman catholic understanding of the natural law, as the Irish bishops saw it, and the duties of the modern state:

It must . . . be kept in mind that 'the moral law', 'the common good of all' and 'the objective moral order' are not derived simply from the teaching of the Catholic Church. They are largely accessible to human reason . . . They oblige all, including non-Catholics, because they proceed from right reason, and not because they are taught by the Catholic Church. They are also an invitation to those who exercise political power to reflect on the nature of man and of human society and in the enactment of laws to eschew the often brilliant attractions of pragmatism, of relativism or of short-term solutions in favour of lasting concern for the common good of all. (Irish Episcopal Conference 1984: 18)

They also expressed reservations over the concept of minority rights: 'Britain, for example, does not allow polygamy even though certain of its citizens accept it from their religious convictions' (Irish Episcopal Conference 1984: 18). However, they did state that, in a united Ireland, Northern protestants would not lose any 'civil and religious rights' (New Ireland Forum 1983–4: xii. 2). This was generally understood by the media to mean that the hierarchy would not oppose the introduction of divorce in a future united Ireland, indeed that they might be prepared to budge on the issue even now. But the bishops' statement does not say so. When forced into a debate on the matter by Seamus Mallon of the SDLP and Senator Mary Robinson, the four church representatives, Bishop Cathal Daly, Bishop O'Mahoney, Bishop Cassidy, and Mrs McAleese, retained the right of the church to enter into public debate on the evils of divorce. Bishop Daly then said:

there must be, there has to be a separation between Church and State and we totally endorse and emphatically reiterate that, but the separation between Church and State does not mean a separation between conscience and the electorate's responsibility in voting. We cannot expect voters to leave their consciences behind them when they go to the polling booth. Inevitably we would expect that those who freely accept the teaching of our Church will vote according to their consciences. This is all we ask of them. But we must have the right to carry out our duty to impart the moral convictions, the moral teaching of our Church to our own members. I should have thought that is part of the reality with which legislators are faced, part of the situation within which they operate. (Cathal Daly, New Ireland Forum 1983–4: xii. 13)

An examination of the documentation shows that there is, in fact, no real change in the bishops' view of the relationship between their perception of the good of the state and public morality, and what was to be *de facto* permissible in the state. There is therefore no elaboration in the forum proceedings on hierarchical directives to their faithful on constitutional matters.[12] Events were to confirm the bishops' implied intent to oppose the granting of divorce as a civil right.

[12] The contents of the present Republic's constitution can be changed by a simple majority of those voting in a plebiscite.

The Divorce Debate

Almost as if to test the political religious alliance of catholic nationalists in the South, the constitutional issue of divorce arose in 1986 as part of Dr Garret FitzGerald's constitutional crusade (see Cooney 1986) to make the Southern state more palatable to Northerners. FitzGerald had been determined as far back as 1964 (1964; 1972) to make constitutional changes to those articles which appeared to alienate Northern protestant opinion. After arriving in office as Taoiseach for a second time in 1983, he proceeded to collect opinions on such issues from various quarters. From the Roman catholic hierarchy, he solicited the response that the church believed in the indissolubility of marriage. Further, 'the Catholic Church does not ask that the law should enshrine any particular provision because it accords with Church teaching'.[13] The statement then went on to oppose the introduction of divorce on the grounds of the protection of 'the common good' by the state. In taking this position, the bishops were also following the lead of Pope John Paul II who, on his visit to Ireland in 1979 had argued: 'Divorce, for whatever reason it is introduced, inevitably becomes easier and easier to obtain and it gradually comes to be accepted as a normal part of life' (1979).

From this FitzGerald could be in little doubt as to the likely course the bishops would take once a referendum campaign got under way. Already, in expectation, the four archbishops had produced a pastoral letter in which they condemned the legalizing of divorce on the grounds of the preservation of 'the common good' and because they feared the tendency that people had to accept as right that which was legally permissible (Irish Episcopal Conference 1985). On the basis of privately commissioned opinion polls showing an increasing majority in favour of some measure of divorce legislation,[14] FitzGerald agreed with his cabinet to proceed with a referendum on whether to change the constitution so as to allow divorce for marriages irretrievably broken down, though only after a period of five years' actual breakdown and legal

---

[13] The quotation is taken from a summary of the Irish Episcopal Conference's statement to Taoiseach, as prepared by the Catholic Press and Information Office, Dublin, and as printed in the *Irish Times*, 8 Apr. 1986.

[14] Forty-nine percent were for some measure of divorce legislation, 36% against, by March 1986: *Irish Times*, 23 Apr. 1986.

separation. The government bill moving the referendum was published in the Dáil on 23 April 1986. Opposing pressure groups were already in existence and preparing for the campaign: in favour, the Divorce Action Group; and against, such groups as Family Solidarity, Women in the Home, and the Knights of Columbanus. These last were soon replaced, on 9 May, by a more specific and highly organized pressure group headed by many who had previously mounted the Pro-Life Anti-abortion Campaign. Of particular importance to the anti-divorce campaign was a lawyer, William Binchy, who had already produced a book in 1984 on the subject. Large publicity campaigns began and the *Irish Times* also lent its weight to the pro-divorce argument, campaigning in its columns until the eve of the referendum in late June.

Though the coalition government were allowing their TDs—the Irish equivalent for MPs—a free vote on the issue, its parties, Fine Gael and Labour, officially sponsored the campaign for the constitutional change. Fianna Fáil appeared to be equally divided on the issue. The new progressive democrats came out in favour of the referendum proposal on 14 May. Of the smaller parties, the Workers' Party was the most decisive in favouring the change. Other public bodies who actively campaigned on behalf of the constitutional amendment were the Irish Congress of Trade Unions and the Irish Council for Civil Liberties. The presbyterian and methodist churches in Ireland, along with the Church of Ireland, welcomed the proposal. They too had been consulted by FitzGerald and had responded favourably.

The four Roman catholic archbishops replied on 28 April that they opposed divorce in general, and particularly the type of what they considered to be unrestricted divorce proposed in the constitutional amendment (*Irish Times*, 28 Apr. 1986). Though in the statement it was mentioned that statistics from other countries were too unreliable to say that pro-divorce legislation increased the instability of marriages and led to an ever-increasing number of breakdowns, Archbishop McNamara was already preaching by 6 May that divorce 'makes stable and permanent marriages more difficult for everyone' (*Irish Times*, 7 May 1986). This was the line already adopted by Binchy (1984) who interpreted world-wide statistics in this direction. Divorce was seen to be harder on the women than the men, who could get out of their family responsibilities so much more easily with divorce than without it. In opposing the arguments marshalled by the anti-divorce groups, the pro-divorce group argued that, despite the existence of divorce legislation in Northern Ireland, there was still a low divorce rate.[15] The period of separation in undefended cases was two years, and five years had to run if one of the parties opposed the divorce. That men in particular do not benefit from the procedures was suggested by the fact that twice as many women as men filed for divorce. It was argued that the pattern in other countries was that laws permitting divorce followed social trends in the numbers of marriage breakdowns rather than vice versa and that the fears of societal breakdown promoted by opponents in Ireland were unfounded in fact.

Despite the position of the church leadership, there were dissenting clerical voices. A priest from County Galway, in a letter to the *Irish Times*, informed that Dr Enda McDonagh, a noted Roman catholic moral theologian, had told priests of the Tuam archdiocese in January that it was possible to make a distinction between the church's teaching on the ideal of marriage and divorce as a civil right. The priest himself, Fr. Standún, intended to vote in favour at the referendum along with 'many clergymen' (*Irish Times*, 10 May 1986). Fr. Pat O'Brien addressed the Divorce Action Group in Galway on 27 May, supporting their campaign on the grounds of religious freedom for minorities (*Irish Times*, 29 May 1986). Professor Sean Freyne of Trinity College, a former Roman catholic priest, published an article in the *Irish Times* favouring the constitutional amendment, and suggesting that the Irish bishops' attitude to marriage, despite some signs to the contrary, was eminently legal and contractual in orientation (*Irish Times*, 5 June 1986).

An opinion poll conducted on 28–9 April and sponsored by the *Irish Times* showed 57 per cent in favour of the divorce amendment with only 7 per cent undecided. But, one week later, this had fallen to 49 per cent in favour (see *Irish Times*, 5 May 1986 and *Sunday Press*, 11 May 1986).

[15] Divorce in Northern Ireland is more difficult to obtain than in Britain, and irretrievable breakdown as grounds for divorce has only been adopted since 1978. For divorce figures on Northern Ireland and their interpretation by the pro-divorce group, see figures supplied by Northern Ireland Court Service to *Irish Times*, 26 May 1986.

Already by 14 May, the Roman catholic hierarchy had responded to the pressure with the production and projected distribution of one million copies of a pastoral letter against divorce (Irish Episcopal Conference 1986). In it they asserted quite clearly that permitting divorce would certainly affect the stability of all Irish marriages because it rendered every Irish marriage dissoluble: 'It is as though the legal availability of divorce builds up a social pressure which, for large numbers of people, becomes stronger than moral or religious resistance' (abridged version, *Irish Times*, 14 May 1986). In effect, they were tending to the view that the very change in law brings about a change in the nature of society and human relationships within it. The bishops also argued that any so-called restricted form of divorce was impossible to maintain in practice and that divorce might solve the partners' problems but only created them for the children. The pro-divorce camp were quick to spot the traditional Irish view that written laws and authoritative words have a radically structuring effect on reality. As Alan Dukes TD argued in the Dáil in response to the pastoral letter:

It is wrong to contend that divorce legislation 'defines' all marriages as being dissoluble. It does no such thing; rather it defines the circumstances and conditions in which a marriage has ceased to be a source of happiness and strength to those involved and may be brought to an end . . . There is no compulsive power in this amendment, nor will supporting legislation contain any obligation on those who do not wish to do so to use the mechanism it will set up. (Dáil debate, 14 May 1986, as reported by Dick Walsh, *Irish Times*, 15 May 1986)

When it fell to Dukes to introduce the second stage of the Bill empowering the referendum, he was forced to address himself specifically to the bishops' arguments in their letter (text, *Irish Times*, 15 May 1986). Similar remarks were made by Senator Mary Robinson, who also challenged the hierarchy's intent in their statement to the New Ireland Forum that the civil rights of Northerners would not be infringed by any pressure from the Roman catholic church in an eventual united Ireland. TDs who opposed the amendment, such as Oliver J. Flanagan on 13 May and Dr Michael Woods on 14 May, began to call on 'the silent majority' to make their opposition to divorce known. The speech of Dr Woods of Fianna Fáil was a particularly important intervention. It indicated that Charles Haughey in particular, and Fianna Fáil in general, were moving behind the anti-divorce lobby. One might be forgiven for thinking that the bishops' letter had something to do with it and that Haughey was intent on constructing an alliance between those heeding the teaching of the hierarchy and the party faithful. In early May, two-thirds of Fianna Fáil TDs from the West had already indicated they would campaign in their constituencies against the amendment (*Irish Times*, 2 May 1986). The Woods' speech also introduced an important new argument which may have had some considerable effect on Irish voters. It suggested that second marriages would substantially infringe the rights of inheritance of the members of the first family. In a society of extremely strong rural, traditional values on rights of inheritance, the suggestions that 'constitutional Frankensteins' would devour 'constitutional orphans' (*Dáil Debates*, 14 May 1986) were clearly calculated to raise a moral panic, having produced the appropriate folk-devils.[16] In the ensuing debate, Alan Shatter, TD of Fine Gael, author of a book on family law, pointed out the better provision being made under the projected divorce law for a first wife than existed at present, and that children of the first marriage would maintain their present rights, though they would have to share with children of the second marriage. There was also much legal debate over the intention not to allow divorce unless adequate and proper provision was made for the dependants. Many critics, including William Binchy, saw this as likely to be an ineffectual provision and subject to the whims of judges. Proponents of the changes did, however, find the prognostications of their adversaries far too pessimistic and polemical. It is interesting to note that the debate centred on the concrete social consequences of divorce legislation rather than on its intrinsic morality. In the Dáil debate of 20 May, Padraig Faulkner, TD of Fianna Fáil argued that the introduction of divorce would indeed end up as divorce on demand anyway, and Charles Haughey indicated his 'personal view' that divorce was bad for 'individual stability' (*Irish Times*, 21 May 1986).

The next stage in clerical campaign strategy was introduced by

[16] 'Societies appear to be subject, every now and then, to periods of moral panic. A condition, episode or group of persons emerges to become defined as a threat to societal values and interests ... the moral barricades are manned by editors, bishops, politicians and other right-thinking people ... Sometimes the panic passes over and is forgotten, except in folk-lore and collective memory; at other times it has more serious and long-lasting repercussions ...' (Cohen 1971: 9).

Archbishop MacNamara of Dublin who instructed the clergy of his diocese to preach on the indissolubility of marriage on the six Sundays from 18 May until the eve of the referendum. He also instructed his parish priests in a letter, leaked to the *Irish Times*, to have the bishops pastoral *Marriage, the Family and Divorce* distributed to every home in the archdiocese along with special bidding prayers invoking divine protection for marriages and the guidance of Holy Spirit 'at this critical time of decision for families and our country' (*Irish Times*, 17 May 1986). An anti-divorce video was released by the Catholic Communications Institute of Ireland and placed on sale in the Veritas Bookshop in Dublin. By the end of the month, it was reported to be 'selling well' (*Irish Times*, 30 May 1985). In an interview with the popular intellectual magazine *Magill* in June, Archbishop MacNamara found he could not accept that the state had the power to determine the meaning and nature of marriage and compared the effects of divorce to the recent Chernobyl nuclear disaster. In pastoral sermons, the archbishop continued to oppose civil divorce, claiming that the weak would suffer as a consequence, along with the well-being of society. He also continued to affirm that the introduction of divorce would make it more difficult for people to lead 'good moral lives'. Divorce has this effect because it suggests that remarriage in the lifetime of one's first partner is socially and even morally acceptable' (*Irish Times*, 6 June 1986; also 31 May 1986; 14 June 1986). This same interpretation appeared in the Irish bishops' next statement in mid-June. Voters had to decide in conscience 'whether other factors outweigh the damage which divorce would certainly cause to individuals, to families, to children and to the whole of society' (*Irish Times*, 13 June 1986). The bishops claimed, however, they were making it clear that people could in conscience vote for the introduction of divorce without incurring moral blame provided they made the decision in a reflective and conscientious way.

Some incidents at local level are worth recording. The parish priest of Brackenstown, Swords, County Dublin, distributed a newsletter at all masses on Ascension Thursday, 22 May—though before he had read Archbishop MacNamara's guidelines on how to conduct the campaign—claiming that no-fault divorce was first introduced by Nazi Germany and that it had since wreaked 'more havoc on the Allied countries than any German army or air force ever did' (*Irish Times*, 28 June 1986). On Sunday, 17 June in Cork a priest was reported to have preached and behaved as follows:

> At the sermon, Father McKenna introduced one of his altar boys, Colm by name, and got him to tell the story about how he and his altar boy teammates had recently beaten another altar boys' team by eight goals to five. When he asked Colm over the microphone how many goals he had scored the boy proudly replied 'Four' . . . Finally, Father McKenna said he had one last question to ask. The question was would Colm like his mother and father to get divorced. Colm said 'No'. (*Irish Times*, 17 May 1986)

Protestant churches were not unaffected by the spirit in which the campaign often appeared to be conducted. Some members of the anti-abortion lobby had apparently been indicating the 'weak' moral stance of protestants on the issue. The Church of Ireland appears to have had this in mind when it invited the well-known English Roman catholic marriage consultant, Dr Jack Dominian, to provide a report on divorce for the Church of Ireland General Synod. As Bishop Empey argued to the synod: 'Somehow we have to nail the lie that permissiveness flows from the Church of Ireland and that morality is the sole possession of but one Church in this land' (*Irish Times*, 22 May 1986). In submitting his report, Dominian stated:

> I have to say that if this room was filled by the Roman Catholic bishops of Ireland and they were pushing me very hard about divorce, I would have to say very clearly that their fear that civil divorce will make marital breakdown worse, or enhance or facilitate it—my answer would be: No, it will make very little difference.
>
> But in justice I think a society needs civil divorce and that the real concern of the Church and the community should be to work at understanding prevention. (*Irish Times*, 22 May 1986)

Dominian's remarks were probably important to the Irish Roman catholic bishops. They had in fact used material gathered by him, and some of his reflections of the sanctity of marriage, in their pastoral letter on marriage in the previous year.

At its annual meeting in Belfast, the Presbyterian General Assembly received a report promoting the amendment as implementing a rightful civil liberty (*Irish Times*, 6 June 1986). Further criticism came from Roman catholics in Britain. An editorial in the *Catholic Herald and Standard* on 25 May—it is distributed in Ireland as *The Standard*—criticized the Irish bishops' apparently low esteem of their laity's ability to remain constant in their marriages. Also a call from the Alliance Party in the North was made by John Cushnahan, party leader. He found it ironic that anyone who supported the Anglo-Irish Agreement could oppose a modest measure to introduce divorce (*Irish Times*, 28 June 1986). Perhaps an important counter to the archbishop of Dublin's interventions was the well-known lay liberal catholic and Head of Information at RTE, Louis McRedmond. He used the columns of the *Irish Times* to inform Roman catholic consciences of permitted interpretations of state divorce on the grounds of religious liberty, interpretations which were diametrically opposed to that of the Irish bishops.

By mid-June, the projection of the opinion polls was already suggesting that the majority favouring a measure of divorce legislation was declining so rapidly that by the time of the poll, the noes would have it. By the week of the referendum, an *Irish Times* opinion poll had a clear and substantial majority against the amendment, 45 per cent for and 55 per cent against (*Irish Times*, 25 June 1986). The referendum proposal was as follows. The present constitution in Article 41.3.2 stated:

No law shall be enacted providing for the grant of a dissolution of marriage.

The proposed change involved the deletion of this and its substitution by the following:

Where, and only where, such court established under this Constitution as may be prescribed by law is satisfied that—

i. a marriage has failed,
ii. the failure has continued for a period of, or periods amounting to, at least five years,
iii. there is no reasonable possibility of reconciliation between the parties to the marriage, and
iv. any other condition prescribed by law has been complied with,

the court may in accordance with law grant a dissolution of the marriage provided that the court is satisfied that adequate and proper provision having regard to the circumstances will be made for any dependant spouse and for any child of or any child who is dependent on either spouse.

The referendum was finally held on 26 June 1986. The appearance of the results of the *Irish Times* opinion poll on the same day predicting the failure of the proposal may well have affected the turn-out. In any event, the constitutional proposal to replace the existing prohibition on divorce by a provision to allow it was rejected. 'Turnout for the poll was 62.5%. Of those voting, 63.3% voted against the proposal' (*Ireland Today*, No. 1030, p. 15). The biggest percentages favouring the proposal occurred in the urban centres, particularly Dublin and Dublin County, where there was a slight margin in favour. But the country areas and particularly the West were overwhelmingly against the proposal.

It would seem that catholic–nationalist morality is to remain a feature of the Southern state and of any future all-Ireland state, if the clerical church's wishes continue to prevail. If an effective will to de-alienate Northern protestants in their attitudes to the republic existed, and if there had been a real change in the Irish catholic hierarchy's position on such matters, then they would indeed have taken positive steps to support FitzGerald's attempt to introduce such constitutional changes. In other words, on the divorce question the Roman catholic clerical leadership were still using their joint authority, which is legitimate in the eyes of the majority in the Southern state, to press into civil legislation their views of public morality, and to use the informed conscience of the faithful as the primary vehicle. They were failing to recognize or were positively discounting the rights of those dissenting from their particular views of public order, and subjecting that dissent to the full coercion of the state. In other words, when the New Ireland Forum was being assured by Bishop Daly of the intention of the Roman catholic church in Ireland to support full civil and religious rights for Northern Ireland protestants, the bishops were effectively reserving to themselves, as a body of luminaries with a direct access to the inner structures of social reality, the right to declare what actually constituted a civil and religious liberty or right and they were doing so on the grounds of what they considered good for society. There can be no question that the bishops are not in any way aware of this arrogation, as it is mediated in consciousness by their belief in, and conceptualization of, a static natural law which is accessible, even if with difficulty, to the conscience of everyman; which same natural law no one should be allowed to violate, even if in error, when that law, if broken, is seen to threaten the very moral fabric of society. The bishops, conceiving of themselves as a body supported by the Holy Spirit in their proclamation of morality and seeing themselves as following the equally and divinely guided line laid down by Pope John-Paul II, were assuming that the opposition to state legislation permitting divorce was of a similar standing and status to the Christian belief that Christian marriage was forever. Therefore, informing consciences on the evil social consequences of divorce was seen by them to be of a similar standing to informing the consciences of catholic lay people on the sanctity and indissolubility of Christian marriage.

It is in this way that the Roman catholic bishops of Ireland, as intellectuals organizing the beliefs of catholic nationalists, continue to be aligned to the forces of conservatism within the alliance, forces which in practice do not recognize important protestant civil rights in Ireland. The alliance is cemented by that traditional political religion, forged in the previous century, which sees the natural law as most accessible to true believers, that it is obligatory to enforce its practice by law, and that those holding other views only have rights to put their views into practice when they are not seen by the bishops to do harm to the social fabric. This monopoly position is therefore totally beyond criticism. One might speculate that an impending solution to the Northern Ireland problem would split off from the clerical leadership those groups for whom the present position is only accepted on pragmatic grounds. Hence the traditional alliance between catholicism and nationalism might be split at the popular basis with the ensuing growth of a degree of anti-clericalism. However, as the very existence of the catholic–nationalist alliance impedes progress towards such a solution, one should perhaps view such a possibility as unreal. The Roman catholic leadership in Ireland has first to change and radically evaluate its monopoly beliefs, thus permitting the dissolution of one of the most antagonistic aspects of catholic–nationalist hegemony.

6
Schooling as Political Religion

Systems and Counter-Systems North and South

INTRODUCTION

A second area where political religion has hegemonic power and structures the popular consciousness throughout Ireland is in the shaping and running of the institutions of education, particularly primary and secondary schools. This chapter outlines the way in which the schooling systems were shaped from the early nineteenth century by a combination of church and state politics, developing in both the North and South as expressions of catholic nationalism and protestant loyalism. They are therefore predominantly sectarian in political religious form, and attempts to modify this sectarian structure are at the moment opposed mainly, though not exclusively, by the Irish Roman catholic clergy. An outline of the church's dominant political theology on the issue is therefore also explored.

EDUCATION AND THE DEVELOPMENT OF THE ALLIANCE BETWEEN CLERGY AND POLITICIANS

A primary education system was established throughout Ireland as early as 1831 (see Akenson 1970; Lyons 1973; Miller 1973). It was basically envisaged to be of national proportions as well as non-denominational, with religious instruction given in separate classes. There was, however, controversy from the start. The Church of Ireland withdrew its schools from the plan which threatened that church's dominance over the primary sector. The presbyterians also had misgivings, and only came back into the system in 1838 when they were practically guaranteed control of their own schools within it. The Church of Ireland rejoined in 1860, because it lacked funds to continue its own system.

Many Roman catholic clerics were unsettled by the plan. The

Christian Brothers, founded in the early part of the century, withdrew from the scheme and unilaterally developed their own system. The concern of Roman catholic clergy about the system was not without grounds. Some feared the use of the schools by proselytizers to indoctrinate the children in protestant principles, especially as the local managers were at first mainly clergy of the established church. There was division among the Roman catholic episcopate as well and as many favoured the system as were against it. The archbishop of Dublin, Daniel Murray, had accepted a position on the National Education Board responsible for the schools. The controversy was so deep that an appeal was made to Rome by the combatants. Archbishop McHale, who opposed the non-denominational system, was hoping for a verdict to support his views. The verdict from Rome, which came in 1841, favoured both and neither. It asked the participants to stop haggling in public and permitted each bishop to take a decision either way for his own diocese (Murphy 1959).

It is noteworthy that at this time (when secularised education was patronised by nearly all the governments of Europe including those that were nominally Catholic) this query was dealt with as a matter of discipline by the Congregation of Propaganda, and not as a matter of doctrine by the Congregation of the Inquisition. (Gaine 1968: 163)

One could perhaps specify further: it was normal for disputes in Ireland to be referred to the Congregation of Propaganda, as Ireland, being part of a protestant state, fell under its jurisdiction. However, it is likely that it was not felt necessary to refer the matter anywhere else as, according to current practice, it *was* a matter of discipline. It is quite clear that Rome did not have a policy on catholic schooling at this time, and certainly not one requiring attendance at such schools as a religious duty. This falls in with Gramsci's understanding of the stages of the development of catholic monopoly. The phase of cultural catholicism was to develop somewhat later.[1]

Over the next fifty years, a significant change in policy took place. Catholics and protestants became more and more segregated by school, as the Roman catholic bishops sought to bind in their flocks from outside influences, on the one hand to protect them

---

[1] As will be seen in Chapter 7 on catholic–protestant marriages, an ethnic catholicism was already being promoted from the political religious centre in Rome.

from proselytism, and on the other to use the schools as a vehicle for the maintenance and development of faith. An indicator of the growing importance of this totalizing view of education was contained in the 1841 rescript from Rome. It announced that, if doctrine as opposed to Bible stories was to be a part of school education, then such religious education could not adopt a common denominator approach between the churches, but had to be total in its presentation of what the Roman catholic church considered to be the truth, otherwise such an approach would be 'dangerous' for the children (Gaine 1968: 164). The implementation of this policy was the work in part of the new Roman episcopal appointees, from Cullen onwards, though the growing hatred of the ascendancy because of the famine and their continued landownership must have provided ample cause for separate schools for the majority of Roman catholic clergy. Already in the 1850s, Roman catholic bishops attacked the setting up of the 'model schools', intended for the pursuit of educational experiments within a multi-denominational setting. In 1860, the rules on the composition of the National Education Board was changed to permit equal protestant and catholic representation. When state-supported secondary education was introduced in 1878, a denominational system was officially set up, though at this time the system was only availed of by the few families who could afford to lose the labour of young teenagers. So extensive became the control of the catholic sector, both primary and secondary, that priests had extensive powers of dismissal over the teachers until the end of the nineteenth century. From 1899, the intention of the clergy to assume and maintain absolute control of their schools led a number of them to refuse to appoint members of the Irish National Teachers' Organization so as to avoid trade union interference. The result was that the union became subordinated for some time to the hierarchy and a breakaway union of protestant teachers was formed. Despite the curtailment of some of the clergy's powers, friction between the teachers' associations, particularly the INTO, and the Roman catholic clergy has remained until this day.

Clergy control of the schools had indeed become a priority for the church leadership. The understanding reached between them and the catholic–nationalist lay leadership resulted in schooling in the future state being allocated to the domain of the church, in exchange for the legitimation of that state by the church leadership.

It was also agreed that supplementary financial aid would come from the state. The agreement came about through a delicate set of potentially dangerous encounters between the Irish party and the Roman catholic bishops during the first two decades of the present century. It proved to be a period of strained relations with the English liberals, who were allies of the Irish party, but who were also in favour of bringing schooling more and more under state rather than church control (Miller 1973: 81–7, 120–38, 268–92). By this time, the schools frequented by Roman catholics in Ireland were also largely under the direct control of the church's clerics. But secularizing moves by the English liberals—over the Education Bills of 1902, 1906, and a private member's Bill of 1911—led the Irish hierarchy to question some of the political intentions of the Irish politicians, who it was feared might sell out the church's interest in schools. According to the Liberal-Irish alliance, the Irish party was supposed to abstain from the 1901 Conservative bill, aimed at bringing denominational schools in England and Wales into the national education system while at the same time absenting them from local government control: that is, giving finance while maintaining the system of denominational clerical control. With the Irish party abstaining, both English and Irish Roman catholic bishops began to pressurize the leader of the Irish party, Redmond, and his associates to support the Conservative bill. Added upset came in the same year from a speech by Dr W. Starkie, the commissioner of National Education in Ireland, who attacked school managers for not being up to their job. The attack was within the context of a speech largely praising the Roman catholic church in Ireland. But the church's clerics still took offence, particularly at the point that local people should be encouraged to take an interest in the schools by having some financial responsibility for them through local government. As priests controlled the local school directly, this was seen to be a direct attack on their role in the school system. Eventually the Irish party was forced to return to Westminster to see to it that the appropriate denominational interests in England were strengthened and to show the Irish political leadership's compliance with the similarly organized Irish system (Miller 1973: 77–85).

This controversy was immediately followed by another. A specially appointed English school inspector reported negatively on certain aspects of Irish schools. He concluded that the division into different schools by denomination, sex, and age resulted in an economically inefficient system. He attributed the inefficiency in part to the managerial system of independent clerical control. Local accountability, state ownership of the properties, and a form of local authority supervision of teaching salaries were advocated as remedies. The Conservative government, through its Irish Office secretary, George Wyndham, then indicated its intention of reforming the Irish system and devolving power to local, lay control, working in partnership with a central government department. The Irish hierarchy immediately campaigned against what was for them a drastic solution. They declared, rightly, that the very power of the clergy in education was being attacked, and, probably wrongly, that the reform of the faults of the present system was only the apparent reason. Possibly in response to the stimuli given by the bishops' public statements, the bishops found that local councils were coming to their aid and support in opposing the proposed educational revolution. The campaign to prevent change of clerical control of the school system was also linked to the contemporary campaign to set up catholic university colleges. The bishops were successful in urging the Irish party to accept the church's programme for Irish education as a whole. Their success was founded on a dual strategy. On the one hand they subscribed to the annual Irish parliamentary party fund and, on the other, they publicly praised the Irish party as the true political representatives of the Irish people and its interests in the British parliament. Even a 'Castle' bishop[2] such as Archbishop Healy, now announced his conversion to the nationalist cause. In other words, support for the education policies of the church were the *quid* the Irish party had to give in order to obtain the *quo* of the bishops' endorsement of the party as the genuine political representatives of the Irish people or nation. The agreement over education was so clear that when the first revolutionary Irish government set itself up in 1918, prior to negotiated independence, there was no Minister for Education. And when the Irish Free State did emerge in 1921, practically the entire school system was under the control of the churches.

---

[2] This time the 'castle' is Dublin castle, seat of British rule in Ireland up to 1922. Sympathizers with British rule were thus appropriately dubbed.

Structures of Ownership and Control: The Cleric–Civil Servant Axis

The present interrelationship between control of the schools and the two alliances varies in each part of Ireland,[3] though one should also bear in mind the additional underpinning of the alliances both North and South by one of the traditional values of bourgeois capitalism, private and selective schooling. Throughout Ireland as a whole, there has generally been an anti-comprehensive ethos in the education system supportive of both class and sectarian divisions in schooling and which is now eroding only in part.

In Northern Ireland, the catholic schools' sector is provided for out of state funds, and remains under the governorship of the local clergy, now assisted by laity and members of the local educational and library boards. Many Roman catholic schools up to 1968 were financed for capital expenditure by the local church as part of the church's determination not to lose control of them. The chairman of the board is usually the parish priest or his curate. With the exception of a small group of public schools, there is the apparently simple distinction between state schools and Roman catholic schools which occurs in Britain. However, it would be naïve to think that control of the state sector, its general ethos, and the teaching of religious education was non-denominational. These schools are protestant, though nothing else is perhaps to be expected when protestant–loyalist teachers teach protestant–loyalist children. Religious education in these schools is officially non-denominational or biblically based and loyalist sentiments are promoted. But religious education has been known to be fundamentalist and in some cases anti-catholic, depending on the teacher. It is important to note that, since the system began in the late 1920s, there has never been a significant move to split up the schools for use by the separate denominations, something which would have been feasible in the larger towns. The Northern public-sector schools come from the partly voluntary system of pre-Partition. In 1928, these schools entered the system of maintained schools, whereby, in exchange for full payment of current costs and 65 per cent of capital expenditure, the former owners and managers, usually churches, were allowed two-thirds represent-

---

[3] For Northern Ireland, see Sutherland, Horgan, and J. F. Fulton in Open University 1973; Murray 1985. Details on mixed-denominational schooling in the South are fruit of the writer's own inquiries.

ation on the local board of management, with the other third coming from the local authority. From 1968, these schools received the same benefits as the Roman cathoiic sector.

Up until the late 1970s, 98 per cent of children raised in Northern Ireland experienced schooling only on their own side of the divide (Darby, *et al.* 1977). Two major exceptions to the general divide were the secondary schools in Fivemiletown and Ballycastle, where shortage of catholic equivalents resulted in *de facto* integration. In 1978, the mould was broken with the setting up of a private interdenominational primary school in Belfast, enabled by a special act of Parliament. This was followed by setting up of a voluntary multi-denominational school at secondary level, Lagan College, in 1981. The school was granted public funding from 1984, since which time a further four similar schools have opened. All of this change was brought about by pressure from the integrated education movement discussed in the next section.

In the Republic, the unwritten church–state accord on education was maintained, and subsequently reinforced, by the emergent catholic–nationalist ethos. The primary state sector, the 'national schools', is paid for by the state, but is effectively under the control of the churches, with a local priest or minister as manager. Since the late 1970s management boards have been set up consisting of three diocesan nominees, including normally the parish priest, two parents, and the head teacher. Until the mid- and late 1970s, there were simply no rules whereby groupings of parents could obtain a state-financed, multi-denominational school, as the only channels of communication in these matters were those between the Department of Education and the relevant diocesan department or other church board. For example, a new housing estate would be built, the diocese would be informed of the development at the planning stage, and the diocesan office would put in for a school, or an extension for an existing school. The arrangement had been unchanged since the early years of the state and had become entirely natural. Similarly for the protestant community: the school plant is owned by the church, or appropriate church body. The main owner of such schools is usually and indirectly the Church of Ireland, and current expenditure and salaries are provided by the state, with the board of governors presided over by the local minister.

The secondary sector in the Republic is hardly so simple in

structure. For present purposes, three items in particular merit comment. Firstly, it was not until the 1930s that a substantial sector of education arose which was not directly under church control. Even then, this was the sector of vocational or technical schools, which were the schools of the politically and economically less influential classes. These were not able to afford education or were unable to pass the necessary scholarship or entrance exams, or they were the sons and daughters of artisans seeking similar type craft training for their offspring. About one-third of pupils have traditionally attended them since that time. These schools come under the control of the Department of Education, through locally appointed committees. They are organizationally secular schools, though some clergy are members of the local governing committees and this has always been a feature. The 'community' or comprehensive-type schools which are beginning to replace vocational and local church schools have a more religious flavour, as they may include the interests of a former convent, which the new school is in part replacing, a former vocational school, and possibly a former diocesan boys' school as well. The churches have never objected to the existence of the vocational schools since their establishment in the early 1930s. It must be added that vocational schools have traditionally had low status. They took what was left after the church schools had creamed off the more academic pupils and the upper classes. Admittedly, the Christian Brothers' schools were open to all, but they tended to be academically inclined in curriculum and ethos. Vocational school pupils in the cities come mainly from the urban working class. In the country, they come largely from rural small holders and farm labourers. In 1978, at least a dozen of these schools were attended by protestants in numbers of between 5–20 per cent of the total (Fulton 1982). It is also true that the religious practice of those who attend vocational schools has been significantly inferior to those who attended catholic secondary schools, at least in the 1970s (Nic Ghiolla Phadraig 1980). It is likely that this is caused not by having less religious education, but by the fact that they tend to be peopled by children of classes who have traditionally fewer links with the church, and include parents making active decisions to keep their children out of the clergy's grasp.

Secondly, there are twenty-eight protestant secondary schools in the Republic. With the steady decline in the Southern protestant population until recently, it was only possible to have government funds for such schools if these schools were reasonably economical. Most of these schools are private, though parents receive assistance with the fees from a committee which is funded directly by the government and which allocates grants by means test. Some of the larger schools, however, are comprehensives, taking in the whole range of abilities, and are government funded for their current expenditure. The consequence of having such large schools with small protestant catchment areas was that those funded from the state had to make up numbers by accepting non-protestants. The Roman catholic element in all the republic's protestant second-level schools varies from one school with 0 per cent to one school with 40 per cent of the total. But some of the headmasters have a deliberate multi-denominational policy in terms of ethos and curriculum and the schools do provide multi-denominational settings which, if researched, could throw light on debates about denominational and shared schools.[4] The current expansion and urban growth in the Southern population has also affected the protestant community, and the increased pressure on schools in urban areas, such as Dublin, is likely to lead to a diminution of the protestant-catholic character of most protestant schools, unless deliberate decisions are taken to preserve the balance.

What emerges from this brief overview is the dominant role of the clergy in Southern education. Clergy have a prime role in setting up schools and a favoured position of direct relationships with the appropriate state institutions. One must remember that it was specifically this issue of clerical power over schools which was the nub of the controversy in the 1900–14 period. In the North, clergy dominate the catholic sector, while a protestant and loyalist culture predominates in the state schools—albeit with a developed secular and liberal aspect in many of them.

THE TOTALIZING EDUCATION SYSTEMS UNDER THREAT: THE INTEGRATED EDUCATION MOVEMENT

The separation of the two opposing alliances in Ireland is thus replicated at the level of schooling. This remains a fact no matter how one seeks to explain or excuse it. It is largely the case that

---

[4] The writer has completed a survey of these schools, but has not yet analysed the data.

catholic nationalists go to one set of schools which have a visible catholic and nationalist ethos, and protestant loyalists go to another, some of which have a less visible protestant ethos, but all of which have a loyalist ethos as well. Some catholic nationalists in the South mix with a proportionally much larger number of Southern protestants who, however, are loyal to the Southern state. Part of the reason why Roman catholics go to their own schools is because they believe they are obliged to do so. The belief is to some degree affected by being told so by their authoritative pastors. It has even been described as a fundamental moral precept by more than one Irish catholic cleric: 'The law of the Church is quite definite, quite universal in this matter. It is that there is an obligation to go to Catholic school . . . It is the same in Northern Ireland as any other part of the world' (Bp. Philbin, Belfast, in an interview on RTE, 3 Oct. 1976, quoted Mark 1979: 10). This prescriptive approach has been repeated as recently as October 1988, when the director of the Council for Catholic Maintained Schools in Northern Ireland said that people who sent their children to integrated schools rather than catholic ones were breaking the law of the church. There is also strong central support from Rome which places catholic education at the centre of its cultural catholicism ideal.

It should be noted that the largest threat to totalizing control of schools in Ireland has so far come from the integrated schooling movement. This movement has a double constituency, one in the Northern, the other in the Southern, state. Their membership, mainly middle-class, have met considerable difficulty in trying to achieve their limited goals—integrated schooling for the children of those parents who wish it—though some of the membership in the early days were aiming at a total integration of the schooling system. Until recently, catholics in the North were practically forbidden by their clergy to attend state schools, exceptions being made in certain outlying areas. Bishop Edward Daly in Derry liberally interpreted the needs of his Roman catholic pupils from the mid-1970s, allowing greater freedom and seeing to it that some alternative religious education was provided. At the same time, the late Bishop Philbin began to refuse the sacrament of Confirmation to Roman catholic children attending state schools. A number of clergy in his diocese refused religious instruction to these children. The reason given was that children would not be religiously and educationally prepared for the reception of Holy Communion and Confirmation because they had not been educated at a catholic school, irrespective of their parents' alternative provisions: in one area, parents had set up their own Sunday schools as an alternative. The measures were felt to be particularly harsh by those catholic parents who were getting involved with the All Children Together movement. Some of its parents had begun sending their children to the less protestant dominated schools within the state system, shortly before the movement was formed. They were doing so on largely political religious grounds. They felt that it was partly their responsibility to bring up their children in an atmosphere of knowledge and understanding of protestants, as they believed part of the difficulties of life in Ulster were caused by this lack of contact. At the same time, they were keen to bring up their children as Roman catholics. All Children Together, or ACT, was formed in the early 1970s. It draws its membership from across the spectrum of religious and non-religious groupings, though they are generally middle-class and mostly committed Christians. They remain dedicated to the task of setting up a group of integrated schools as a sector of education complementary to the existing system. An appeal made to Rome in 1977 by the Roman catholic parents concerned with the Confirmation ban appeared to have been answered in their favour, as Bishop Philbin changed his pastoral policy and these children were allowed to go forward for confirmation in 1978–9. A notable achievement of ACT has been the passing of the Education (Northern Ireland) Act of 1978, permitting the establishment of multi-denominational schools where such is desired by adequate numbers of parents. The Act is an enabling one and does not affect the existing system unless the leadership of the churches wish to co-operate in a reshaping of the system or of schools in any particular area. In addition, such schools have to be self-financing until they have an adequate complement of pupils.

In the Republic, as in the North, a parents' movement to set up integrated schools was started around the same time. It grew up originally in the Southern middle-class suburb of Dalkey, County Dublin, and around the Church of Ireland national school there. A dispute with the local rector and manager led to a group of parents withdrawing their support from the school and coming together to form the Dalkey School Project. The organization claimed membership from all parts of the South, and its membership contained education specialists, lawyers, people from the media, and civil servants, at least one of whom knew something of the existing channels of power and communication between church and state in the national school system. Energy was directed mainly to the primary sector, where, until recently, it was impossible to have anything but a church-sponsored school if it was to be funded by the state. In 1978, on the basis of their professional knowledge of schooling and civil administration, the organization succeeded in obtaining the necessary government permission to open in Dalkey a multi-denominational school governed principally by the parents, the mode of religious instruction to be determined by them. It opened the following September. Another project at Marley Grange in South Dublin was less successful, due in part to the Church of Ireland deciding to support its own primary school in the area, and in part to the open opposition of the local Roman catholic clergy.

At Firhouse, in South-West Dublin, a third project was in the making in 1975. In a survey with a 75 per cent response rate undertaken by the residents' association on its new housing estate, some two-thirds of the respondents favoured the setting up of a multi-denominational school. The present writer, who was not connected with the survey, estimated that there was about a 15 per cent non-Roman catholic element on the estate at the time. There seemed to be the fairly uncomplicated desire among the community to see all the children attend the same school, as an expression of relationships existing within the community. It is likely that children have the important function in a new housing area of bringing parents and neighbours together, particularly when there is a lack of other social amenities—a characteristic of new housing areas in and around Dublin in the 1970s. The children probably appear as a source from which to develop new relationships and the immediate perception is to translate this experience into scholastic terms. The absence of a non-Roman catholic school in the immediate area and the naïve belief that they were empowered in some way to have a say in what type of school should appear on their housing estate—there was a small Roman catholic school which was to be expanded to cater for the growth of the population—may have sharpened catholic parents' interest in having an integrated school. There were at this point no apparent difficulties experienced between the majority of estate members and no significant source of conflict, that is until the question of a multi-denominational school itself became such a source.

The two Roman catholic priests, who were in all respects dedicated pastors and much liked by many in the local community, immediately opposed the idea, preaching against it at Sunday masses in the local convent and the school hall. The burden of the message was that good catholic parents sent their children to catholic schools. The curate added to this that those promoting the integrated project were in fact promoting secularism. The fact that a prominent member of the current community council and an integrated education supporter was a member of official Sinn Fein, the Workers' party, appeared to figure in the reasoning, as this party has always been suspected to be an anti-clerical and secularist force. In residents' association meetings, the clergy's point of view received vocal support from one or two members of the older village community which preceded the housing estate. The total of those favouring denominational schooling in the initial survey was less than one-third. But, since the open opposition of the clergy, many of those attracted by the original idea were clearly dissuaded and the majority opposition dwindled away. The clergy had informed the people's conscience on the basis of what they considered to be the essential religious interests of their flock. Alternatively, from another perspective, the clergy had acted to preserve the old divisions. Unfortunately, the one is an inevitable corollary of the other.

ROMAN CATHOLIC VERSUS INTEGRATED SCHOOLING: INTERPRETING REALITY

The clearest opposition to the achievement of a measure of integration has been from the Roman catholic church. But where other churches have owned their own schools, as does the Church of Ireland in the South, a similar position has been adopted, though the reasons for the position have been on different grounds. A proposal before the Church of Ireland synod in May 1979 to support interdenominational education was defeated, despite pressure from its ecumenical lobby.[5] The main grounds appeared to be

---

[5] See Elliott 1978. If the Roman catholic clergy ever softens its line on integration, Northern protestants are likely to become more oppositional to integration also; see Murray's findings (1985).

the danger which it might pose to the small Southern protestant minority by encouraging mixed marriages. However, no evidence favouring such a contention has ever been produced. Harry MacAdoo, then the Church of Ireland archbishop of Dublin, on several occasions put forward the argument that Southern protestants were a minority group with their own culture and traditions, and thus deserved to have schools for themselves in order to hand on their own traditions. When considering the merit of this argument, one does have to bear in mind the particularly precarious nature of the Southern protestant grouping.

Protestant opposition to integrated schooling is sometimes heard in the North, particularly from the fundamentalist camp, who not only fear catholic infiltration of state schools but are opposed to anything other than Bible protestantism in religious education. But usually the opposite is the case and popular protestant support for integration is expressed. Perhaps if integrated schooling were more widespread in the North, more protestant voices would be heard against it, particularly if there was a possibility that a nun or priest might show up as a teacher in such an integrated school. There is also evidence of hardline evangelical indoctrination by some protestant teachers in state schools. In addition, Roman catholic schools in Ireland, in continuity with the belief and practice of Roman catholics in a number of other countries, have a different concept and practice of teaching religion in the schools from that of the majority of protestant traditions. This ranges from the presence of religious symbols in the classrooms and corridors of the schools to religious services such as mass or the saying of the rosary, and finally to the specific form of religious education which is praxis oriented. Learning, singing, and praying are rolled up into one: catechesis, as opposed to dialogue or concerned interest for religious matters. Children are prepared in the classroom for their first communion, confession, and confirmation.

The Controversy over the Reality of Schooling

Extensive public debate on integrated and segregated education has taken place throughout Ireland.[6] Bishop Cahal Daly has argued

[6] There have been numerous meetings of academics and notables, and sponsored seminars. A number of papers are available on the topic. The writer participated in several such meetings especially between 1976 and 1979. There has also been a considerable amount of material published in books, pamphlets, reports,

that the debate on multi-denominational schools has effectively taken people's attention away from the central issues in Northern Ireland, namely social injustice, fundamental political inequality, and the violence. In reaction to this, one can say that the very way in which the church leaders view the subject is itself revealing of significant mediations between religion, state, and national ideology in the institutions of schooling. In the debates over the last twenty years there has been considerable interplay of theological and sociological reasoning, with social scientists also contributing to papers and discussions. What has been seen as an aspect of the Roman catholic intellectual opposition to divorce in Chapter 5 can also be recognized as a feature of the defence of catholic schools too: the opposition contains an interpretation of the moral nature of contemporary society and of what happens to catholics who are not to some degree protected from it. In this writer's view, the arguments against multi-denominational education and in favour of catholic schools support an overall political theology which is strongly based on authority within the church. Hierarchy and clergy are seen almost to monopolize the teaching and wisdom about life. They are both directors of souls and minders of the public conscience. Put together with a faith vision of the contemporary social world as inherently secularist and radically subversive of Roman catholic values, the combination is significantly reactionary and ethnocentric.

The aspects of this political theology will be discussed later. First it is necessary to consider the arguments used in the public debates. While the arguments of the defenders of catholic schools have a theological basis, those of the promoters of integrated schooling have been founded on what they see to be the results of the present system, namely a violent and destructive society. But there has been some attempt also to undercut the basic level of the defenders' case. It has been argued that catholic schools do not do the job for which they were set up, that is educate Roman catholics sufficiently to keep them in the church. Irish bishops have been well aware of Spencer's (1968) research. Using traditional measures of religiosity, he has pointed out the apparent failure of English catholic schools to produce better catholics and fewer ex-catholics than state or

and newspapers: see in particular Ecumenical Notes 1976, and, more recently, *Irish Times* for 1986. The writer has also used notes from talks and sermons on the subject between 1974 to 1979.

other schools, and has inferred the likelihood of the same for Irish schools. Spencer's argument has been banded about from time to time. This is not unnatural, as in the 1970s and early 1980s Spencer was at the Queen's University, Belfast, and an active member of the All Children Together Movement. Support for this line of approach has been found in Greeley and Rossi's (1966) survey in the USA on the effectiveness of catholic schools. Based on statistical data for Roman catholics educated in both Roman catholic and other schools, Greeley and Rossi noted as one of their findings that there was little difference in adult religious practice between the two groups. There was a group among former Roman catholic school pupils who exhibited a more intense level of practice, but these were also children of fervent Roman catholics. The causal inference adopted by Greeley and Rossi was that Roman catholic schools only had a clearly identifiable and reinforcing effect on the belief and practice of this group. Clearly, Greeley and Rossi's findings could be used both ways, and in fact both camps in the Irish debate have found it supportive of their case.

In a similar survey undertaken in 1974, Greeley found the effectiveness of catholic schools either the same or slightly increased. In the decline the church had experienced in the intervening years, fewer catholic school children had opted out. In addition, whether or not one had attended a catholic school may have had little effect on sexual values or mass attendance but it did increase catholic activism and racial tolerance (Greeley, McCready, and McCourt 1976). In the light of W. McCready's findings in another study that the religious behaviour of the father had a greater impact on children, Greeley considered the fact that catholic schools were having an increasing influence on men to be of significant importance for the future of Roman catholicism in the USA (1976: 173–5). Clearly, the follow-up study was more favourable to the defenders of catholic schools than to opponents. However, Greeley and Rossi's work also has been used to uphold the defenders' argument that, though Roman catholic schools varied in their efficiency, they were certainly better than any possible alternative in terms of producing good Roman catholics.

Thus, both sides have tended to use social scientific research to make statements in general about catholic schools which have been drawn from other countries and, therefore, beyond the cultural confines within which Irish catholic schools exist. This is something for which surveys, including that of Greeley and Rossi, are not and cannot be intended.[7]

Another point worth bearing in mind is the very generalizing character of Greeley and Rossi's studies. There was nothing in the studies on the differences between individual catholic schools, either in terms of pupil experiences or the varieties of educational practice and personnel in each school. A black-box approach to the schooling process was used with no direct investigation of processes within the schools. Finally, while it could be argued that Greeley and Rossi tell us something of catholic versus state schooling in the US, they have nothing to tell us about catholic versus Christian, multi-denominational schools, which is what most of the argument in Ireland is about.

But the central and substantive argument that defenders of Roman catholic schools have had to deal with is that the dual school system in Northern Ireland encourages, supports, or at least reinforces the sectarian divide, and that it forms part of the vicious circle maintaining conflict in Ulster. It has been argued that one could help to break that circle by integrating the school system (Fraser 1974; Heskin 1980). A small number of empirical studies actually done in Northern Ireland have been invoked in the course of the debate.[8] Rose (1971) pointed out that those with a mixed schooling background showed marginally less extreme political views than those educated separately. Unfortunately, Rose did not elaborate on the nature of such mixed schooling experiences. They

---

[7] Survey results are only sociological by implication. They can only be referred to another population in a different cultural and historical milieu on the basis of a clearly argued theory. Indices are always abstractions from given socio-historical wholes and are therefore essentially relational. They cannot be extrapolated from their context without due care and attention. Social institutions are rarely identical cross-culturally. In the case of Roman catholic schools, there are a variety of such organizations around the world and each exists also with a different national framework of cultural and political relations. Such schools in Britain are largely a part of public sector education, and those which are private cater for the upper classes, and operate significantly favourable staff–student ratios. In the USA, catholic schools are entirely private and their children represent a disproportionate number of the middle classes. Even if catholic schools in the USA, after controlling for class, ethnic background, and private or public funding, may not as a whole have reinforced prejudice or divisions within the social order, this does not mean that catholic schools in Ireland would behave in the same way. That would remain to be shown by investigations in Northern Ireland itself.

[8] They are Rose's (1971) survey of attitudes prior to 'the Troubles' and surveys in schools by Russell (1972) and Salters (1970), and Murray's participant study (1983; 1985).

could even have proven to be the negative experience of small ethnic enclaves defending themselves in hostile environments, but even this viewpoint can only be speculation. Murray (1983; 1985) advised not to expect too much of schools on the basis of his own in-depth research into two primary schools, one from each Northern group. He has also sought support for this view from studies of separate and mixed schooling of different ethnic groups in other countries. Such studies, he argued, have shown that joint schooling is not necessarily successful in lowering prejudice. Consequently, one has to improve relationships within Northern Ireland without trying to merge the entire school system. However, curricular changes, joint teaching projects, and overall greater contact were seen necessary.

Roman catholic church leaders have used additional local studies to refute arguments which treat Roman catholic schools as part of the Northern problem. They have interpreted Salters's study of Belfast secondary schools as showing catholics to be less prejudiced than protestants. Even if this were so, the implication has been made by catholic school supporters that catholic schools have not promoted or reinforced any measure of prejudice, and that no other possible type of school could improve on them (Daly 1975). Unfortunately, the necessary link to warrant this second conclusion—a control group of catholics in a multi-denominational situation—was simply not available to Salters, and so the argument is void. Use has also been made of Russell's work on the teaching of civics in Northern Ireland schools. The work led Russell himself to conclude that the teaching of the subject had no effect whatsoever on pupils' political outlook. This can be taken either as a criticism of the actual teaching of the subject or as a remark on the ineffectiveness of schooling on such attitudes in the Northern Ireland context. Russell tended to support the latter view. This has been taken to mean that teaching children together with the aid of a definite programme to reduce prejudice would not work, thus invalidating any proposed integration of schooling. This, of course would have to be seen. But both Salters's and Russell's work in Northern Ireland schools have been used by catholic schools supporters to aid a general principle, also derived from the work of some sociologists of education: namely that schools have little impact on children in terms of their basic values anyway, and one should recognize parents and community environment for the real source of these. The tradition in educational studies which seems to aid such views stresses that you cannot change society with the school. Schools cannot have a real effect on children's basic values because these have been instilled earlier by the process of primary socialization in the home and in the wider social community. Of course, other studies could just as easily be cited to state the opposite. Schooling, according to Eggleston (1974) and Shipman (1972), has a definite effect on human behaviour and values, and, according to Bronfenbrenner (1972) can even enter quite profoundly into the socialization process. The debate is now largely 'old hat' in the sociology of education and there is a general consensus that schools do indeed impart values, but these are usually modified or developed versions of already existing dominant values of the society of which they form a part, including a number of parental values. Needless to say, any such denial of this would be a denial of the power of education as a whole to mature individuals in their intellectual and moral perceptions. Again, according to Bernstein (1975), Ball (1981), and many more, schooling can have opposite effects on children according to whether children have been socialized into working-class or middle-class aspirations, and according to the actual way the schooling process is structured by the state, the teachers, and the subject-disciplines. At the same time, sociologists of education generally hold to the view that schools can have an important role in both conservation and social change. Consequently, the intricate role they play in the North or South of Ireland is for the investigator to find out.

The Theological Component in the Ideologies

The sources allow one to construct an outline of the dominant Roman catholic ideology on schooling in Ireland (particularly: Guild of Catholic Teachers 1971; Conway 1970; Daly 1978; Faul 1977; Philbin 1975). The strongest form of the argument for catholic schools runs as follows. A school controlled by catholic clergy, religious, or practising and committed lay catholics, a school which operates a thoroughly catholic, religious education syllabus taught by dedicated and religiously committed teachers, and a school which is permeated by a thoroughly catholic ethos is the only way to educate Roman catholic children. In all possible situations, the establishment of a Roman catholic school is an obligation of conscience of basic priority and stemming from the mission of the church. Any type of religious education which does not presume a commitment to Roman catholicism as a community and as a faith, and is not catechetical in nature will result in a watering down of that faith and a drifting away from that community. If religious education does not consist in handing on the faith, it will be destructive of the faith itself. Integrated schooling in Ireland must therefore be largely rejected. Some have even continued the argument with the view that all education is ideological and the main problem is to teach the right ideology.

The argument as a whole is also linked to a number of other religious political perceptions, particularly the primacy of the family. The school has been seen as having to support the parents in their own work of education and the values of the home environment. The school cannot be allowed to confuse the child by presenting values which conflict with those of the parents. At conferences and talks on the topic, this view was found to be more explicitly defined: the school should have a denominational atmosphere which reflects the denominational atmosphere of the home. The view is shared by some protestant as well as Roman catholic clergy (e.g. Church of Ireland Bp. McAdoo, *Irish Independent*, 5 Apr. 1971; Card. Conway 1970).

At this point it should be said that if all education were ideological and religious education only a catechesis in a narrow sense, then indoctrination would be its sole outcome and what purports to be education a sham. No genuine critique of ideology appears to be taking place when such statements are made. Fortunately, Irish catechetics experts are aware of the problem (e.g. Brennan 1974; Greer 1975),[9] but they are not the persons in positions of political religious power. In addition, the non-fundamentalist protestant tradition of a liberal religious education is based on principles which are still opposed by many Roman catholics as well as by protestant fundamentalists on the grounds that they imply a moral and religious relativism. This is also why religious education in catholic schools is usually an extension of what goes on inside the Roman catholic home and church, and has

---

[9] In an otherwise excellent study, Murray (1985) implies that the beliefs on schooling of Ulster catholic clergy are the only possible catholic views on the subject. In fact they represent a particular catholic interpretation not shared by all Roman catholics, bishops included.

a cultic dimension. In this view, there is no place for a formal religious education without the faith experience being totally bound up in it, or even for an encounter with the religious beliefs and practices of other churches or religions. This lack of perception of such alternatives is a characteristic of hardline adhesion to the catholic school principle, despite the attempts by catechetics experts to point out these other real possibilities.

Also, the relative importance of the rights of the family is a recent argument. As Gaine (1968) points out, the legitimations of Roman catholic schools in both Britain and Ireland have changed over the years. In the nineteenth century, they were defended on the grounds that only priests were called to religious instruction. At a later stage, the argument that the use of the state authorized version of the Bible in the classroom was not acceptable to the Roman catholic church predominated. Finally, the argument that proselytism would take place in interdenominational schools came to the fore. In the 1970s, what was being observed by this writer in the current debate as the main source of legitimation in Ireland, was the argument that parents have a right to the kind of school and ethos which they desire for their children. But as the church, seen to be operating in its authoritative hierarchy, tells parents that they must want to send their children to Roman catholic schools because they are obliged to do so by their faith, it follows, in the logic of the argument, that Roman catholic parents do want to send their children to Roman catholic schools, or will want to when their consciences have been appropriately informed.

The variety of arguments used by churchmen over the 150 years of the debate on the church control of schools appears to have been an attempt to cope with the enormous ongoing social and industrial transformation and to comprehend the mission of the church in such rapidly developing circumstances. For the high clergy, the transformation required a new form of monopoly power appropriate to match that of the state, and clearly the issue of the rights of civil and 'natural' institutions *vis-à-vis* the new state form was and remains a major issue. However, that civil democracy itself might be seen as a definite advance in terms of religious development itself was never really confronted.

The second element in this theology is a faith reading of the nature of the contemporary world: the Irish Roman catholic clergy have seen the major enemy in the new social form as secularism. As has already been seen, since the nineteenth century Irish Roman catholic bishops have been mainly concerned with the preservation of the church's position within the prevailing social frame of reference. By the 1950s, they felt confident enough to say that a 'catholic social order' had been achieved in Ireland, though it is not clear from the statement whether Northern Ireland was included in this (Maynooth Synod 1956). The conservative attitude which this view exposes is continued today in the belief that a post-Christian society has arrived in Ireland. The demise of Christianity appears to have been so keenly felt that the attempt by the joint churches' Ballymascanlon working party IV (1976) to redress the balance and to indicate positive values in secularization appears to have been largely abandoned as a futile academic exercise.

It does seem that, for the bishops, secularism—a state of mind in which God has disappeared altogether as a meaningful point of reference—has been the great threat. Secularization, or the loss of control by the church of various areas of social life, has appeared to be the way of such secularism. This has been seen to apply especially to education, that area of society in which the church had direct control.

The conviction has appeared among the majority of bishops, at least among all who spoke out on the issue, that for the clergy to lose control of management—the actual appointment of principals and teachers, the right to oversee their conduct, and the right to pursue Christian initiation or preparation for the sacraments in the classroom context—represents the thin end of the wedge of secularism.[10] The state is still seen to be capable of pursuing a policy of gradually excluding religious education from the syllabus or declaring it neutral ground and one would end up in Ireland with secular schools populated by increasingly secular children. As a consequence, the integrated schooling movement was frequently seen as a vehicle of secularism. It should be stressed that both the Northern movement for integrated schooling, ACT, and the Southern, the Dalkey School Project, have a large number of committed Christians among the rank and file and among the officers of the associations. One could almost say they predominate.

---

[10] Significantly, the phrase 'thin end of the wedge of secularism' came back into vogue with the anti-divorce campaign also.

Power of the Religion of the Intellectuals over Popular Religion

It is interesting to note that the theological argument under review only appears to have been pursued by national episcopal conferences where the church has inherited a system of Roman catholic schools, or where it has built them up when no other schooling was available. Even then, where the state has taken over the system, church leaders appear to have adapted their demands to the new situation and, in some cases, an interdenominational school syllabus has been put into operation, as for example in Zambia, Tanzania, and Kenya (Smith 1980; Ecumenical Notes 1976). In such countries, there has been no further demand for separate schooling but the church appears to have found the new system adequate to its mission. There are some indications at least that these bishops see the issue to be more a matter of discipline than of faith, though of course it may well be a case of bishops making the best of an inferior situation to the one desired.

Clearly, the Irish catholic leadership have revealed in the debates their perceptions of how the church should fulfil its mission. There appears to have been an assumption that the church, in its authority structure, has already found the correct ways to exercise power in modern Ireland. One of these is to influence the people directly through clerical control of the school system, within which the church hopes to raise a new generation of Roman catholics, particularly the leadership in society. In this, the clergy are still following the path of cultural catholicism. They also hope to influence the faithful through direct action on the state as an episcopal pressure group: no longer through hidden channels but by policy statements at appropriate times, such as on the occasion of crucial voting in the Dáil. But, as has been seen, the emphasis appears recently to have shifted to shaping, with authority, the consciousness of the laity particularly on constitutional matters. Notwithstanding, in the context of the school direct clerical authority is still sought. In the Republic, three of the six members of the primary school board are episcopal appointees and in most new community schools a majority of the bishop's appointees still pertains. During the period 1978–9, there was considerable resistance from the religious congregations involved in schooling to plans by the government to give them a minority role in the management and organization of new community schools into which their own schools were about to be absorbed. From these developments, it is clear that bishops and clergy still believe that the laity must remain in a subordinate role where schools are concerned. The clergy cannot trust them with such a serious matter as they see schools too bound up with the issue of authority in the church. The laity are seen to be inherently unstable vessels of the church's mission.

One of the results of the events and arguments on integrated education, particularly the felt determination of the Roman catholic clergy to maintain control of schooling, is that a real opposition to this mode of clerical–lay power relationship is emerging in Ireland, a form of pro-religious anti-clericalism. Anti-clericalism is therefore not just a phenomenon related to persons unconnected with the churches. Members of the Roman catholic church are themselves involved, in their roles as parents, politicians, and bearers of overall human responsibilities as members of society. In general, these opponents appear to be members of the more articulate middle classes, and mainly urban dwellers who are caught up in the general administrative and professional life of contemporary industrial society. It is an aspect of the religion of the middle-income brackets and city suburbs. As politicians and civil servants, they are seeking more control over their state's expenditure on schools, motivated both by financial stringency and moral obligations to the taxpayer to account for actual expenditure in the area of education. An equally pressing demand would appear to be the need for the Republic of Ireland to match other advanced industrial societies in the quality of education (see OECD Reports 1965; 1969). The result is an attempt to achieve greater public accountability by extending state control into school management, either through local government, as in the case of vocational schooling, or through the Department of Education, as in the case of community schooling.

The effect of the clergy's position on the laity appears twofold. On the one hand, there is a closure of the mental and moral attitudes of the laity to the issue of shared schools. But on the other, an influential group of laity has come to experience the feeling of being left more and more out in the cold. In the South, the churchmen have never consulted the laity in any given community before embarking on building a new school. Even the state has had until recently, no official intention of considering wishes expressed by local community groups on such issues and has continued to work the traditional system of direct contact with the church authorities. Hence there has developed a form of alienation of Roman catholic laity of the new urban middle-classes from both the clergy, as well as the state, though to a lesser degree. The state, however, might well be affected in its actions in future by local communities through the exercise of election rights. The community might be encouraged by pressure groups within political parties to pursue multi-denominational schooling. But, particularly in the light of developments in other areas such as the national referendums on abortion and divorce, the souring relationship with the clergy holds forth no immediate prospect of resolution.

In old established communities, the local laity see no need for changes and do not question the church's preserve in education. The clergy and religious orders who are actually running schools find a substantial support from Irish catholic laity who have an undoubted attachment to the schools of religious orders. This becomes clear from the joining of forces which takes place in the resistance to comprehensive and community schooling. This tradition among the middle classes of private and semi-private schooling, around which identities have grown over the years, would appear to be a powerful element in resistance to government-sponsored changes in the educational system North and South. But from such motivations must not be excluded certain class interests in preserving the privilege of better educational opportunity for those who can pay for it—better it must be said than the opportunity offered to the children of less fortunate parents. The opposition of such movements as ACT and the Dalkey School Project, represents the growth of a middle-class movement in opposition to the lure of the twin blocs in Ireland. In the Republic, it mainly represents a modification internal to the catholic–nationalist alliance. But what still predominates is a religious culture diffused by the clergy which makes a substantial input to the structure of the divisions between catholic nationalists and protestant loyalists, a division of which the educational institution itself has come to be a part.

CONCLUSION

Given the seemingly rock-solid faith in catholic schools which the high clergy and their supporters have, one has to ask the question why they should bother to defend them so vigorously. Two immediate answers can be provided. Failure to defend them publicly and on grounds evident to all, including secular politicians, might lead to the loss of catholic schools or ending up with what the clerical leadership would find to be a watered-down version of them. Again, the linking of catholic schools to 'the Troubles' might cause scandal to the faithful who might then begin to consider alternative schooling unless persuaded otherwise.

But there are other levels of motivation which should be considered. To lose dominant control of the schools would require an entire rethink of mission and pastoral strategy. Though the loss in real terms might mean greater participation of their own laity in running all the schools, and some sharing in integrated projects in those middle-class areas where such schools seemed feasible, considerable dialogue and agreement with protestants on the management, ethos, and curriculum contents of such schools would still be required. Alternative arrangements would also have to be made for Christian initiation into the sacraments, probably outside school hours and involving parents and others within the parish structure in an unpaid capacity. Quite a change in world view is required in such a reorganization, not the least of which would be having to think through one's principles in the field of religious education.

At a deeper level, another motivation occurs. As has been seen in the previous chapter, Roman catholic teaching is seen to be very close to the natural moral order and to be supportive of that order in society. The major scandal—to clerics themselves as well as to laity—would be to discover that one of the very things to which the church is actively committed violates in some fundamental way the common good and the moral order. This takes us back to Chapter 1, which mentioned the problem some Roman catholics have in finding sin within the church, or within at least what can be considered one of its current constitutive ideological practices.[11]

There remains an even more radical motivation. What has grown up in Ireland among Roman catholics—and in many other parts of the Roman catholic world—is an ethnic sense of identity, rooted in a particular interpretation of Roman catholicism and the modern

---

[11] The term 'constitutive ideological practices' is derived from Gramsci and made popular by Hall (1986). It stresses that the most significant social element of beliefs and cultures are those which enter into and flow from what one does in society, from working on the factory floor to running a government.

world. There is a sense of separateness which is based on a concept of religious superiority and on the sinful and empty nature of the surrounding world, encapsulated in the concept of secularism as negation of the truth of religion. This sense of separateness is fed by an understanding of the wholeness and truthfulness of the faith one holds and the radically flawed nature of all alternatives. Other churches are considered less than such, despite Vatican II teaching to the contrary. There is therefore nothing really to be learnt from others in the surrounding society and the best thing is to let it be, and not to mix with its inhabitants to any extent which would lead to a weakening of this absolute faith. In the context of a society shared with protestants—some of whom hold a similar view—the effect is quite devastating. It firms up the separateness of the dual alliance structures—indeed it constitutes one of the principal components in the separating process itself, justifying and promoting the maintenance of the void. As will now be seen in Chapter 7, the promotion of such separation is also firmed up at the margins, by inhibiting the mixing of the two alliances at the root, in the field of marriage and procreation.

7
The Religious Boundaries of Group Reproduction

Catholic–Protestant Marriages North and South

EXTENT AND DEMOGRAPHIC IMPORTANCE

A third area where political religion exercises hegemonic power while at the same time shaping the catholic–nationalist and protestant–loyalist alliances, an area where church law and ethnic divisions combine to draw strict sexual and procreative boundaries around them, is the area of intermarriage between protestant and catholic, or 'mixed marriage', to use the local and old ecclesial terminology. It will be shown how theological doctrines and church discipline have combined in the past to brand mixed marriage as highly dangerous, morally suspicious, and frequently sinful; and that this ecclesial perception still matches the religion of the people and common-sense perceptions of 'the other side'. In this way, an infrequent occurrence—mixed marriage—comes to have major significance in group reinforcement. The practice is revealing of the relationships between the two alliances. It challenges bloc solidarity and infringes cultural boundaries. To use an analogy from Mary Douglas, a pollution occurs in the lives of both groups (1971: 136).[1] Occasions of deviance at the boundaries of group perception are ones which are heavily laden with sin and impurity, as Douglas has shown in her now famous exegesis of the 'Abominations of Leviticus'. More importantly, the issue of mixed marriage reveals the extent of church involvement in the processes of group reproduction and the maintenance of sexual and procreative boundaries between the two sides.

It is still quite difficult to construct accurate figures on the extent of intermarriage in Ireland. Valuable information has been provided by Lee (1981: 71–96) and, more recently, by Compton and

[1] The concept has been applied to the topic in Fulton 1975: 167.

Coward (1989: 182–97) for Northern Ireland (see also Table 7.1). All admit to the difficulty of calculating the number of couples in mixed marriages who have left Northern Ireland or even Ireland altogether. Lee's figures include converts from one church to another, whereas Compton and Coward identify both present religious identity and religion of origin. Compton and Coward's figures are the more impressive, because they are based on the Northern Ireland Fertility Survey of 1983. As it turns out, Lee's estimate of the intermarriage rate of those living in Northern Ireland is somewhat lower than the one provided by Compton and Coward, but was the only reliable source until the findings of the fertility survey were produced. In any event, the rate is exceedingly small compared with England and Wales, where the rate for Roman catholics marrying others runs at about 67 per cent of all Roman catholic marriages. In Northern Ireland, those born catholics intermarry at the rate of 6 per cent, and those who remain catholics after marriage intermarry at the rate of 3.6 per cent. In Britain, Roman catholics form about 10 per cent of the population as a whole (Hornsby-Smith 1987: 43, 92–3). Being a minority, however, does not automatically lead to mixed marriages. In Northern Ireland, at least up until the 1970s, methodists married

TABLE 7.1. Mixed Marriages in Northern Ireland as a Percentage of Total Number of Marriages (sample number)

Year	Total	Mixed	% Mixed
1978–82	462	45	9.7
1973–77	508	36	7.1
1968–72	563	28	5.0
1963–67	424	26	6.1
1958–62	349	19	5.4
1953–57	280	8	2.9
1948–52	203	8	3.9
1943–47	77	1	1.3
Total[a]	2,923	176	6.0

[a] Excluding missing values.

Source: Compton and Coward 1989: 186. Figures are drawn from the Northern Ireland Fertility Survey of 1983.

almost exclusively within their own denomination (Lee 1981: 74). Overall, the relatively low rate of mixed marriages is likely to be controlled by the strength of solidarity within the two alliances in Ulster.

As is shown in Table 7.1, the figures provided by Compton and Coward, the rate of catholic–protestant marriages is on the increase, despite 'the Troubles' and despite the dip in the rate at the height of the violence in 1968–72. In addition there is a higher rate for the under-20 and over-24 age groups than for the 20- to 24-year-olds: 4.4 per cent for the 20- to 24-year olds, 7.8 per cent for the under-20s, and from 7.3 per cent for the 25 to 29 age group to 13.6 per cent for those over 34. The data suggest two opposing factors at work. On the one hand, the contacts and culture of modern society and changes in attitudes to intermarriage on the part of the couples involved are pushing up the rate of catholic–protestant marriages. On the other hand, pressure from kin and community is at its greatest on those most likely to have families, namely the young adults. The high rate of intermarriage for the under-20s is explained by the high level of pre-nuptial conceptions in the group of mixed-married couples.

Compton and Coward reiterate two important points made by Lee. There are more working-class than middle-class intermarriages, and intermarriages in rural areas appear to be as numerous as in Belfast and other areas of urban growth. The main exceptions are the North-East, where the intermarriage rate is relatively high at 13 per cent, and the area of South Armagh and South Down, where it is only 2 per cent. The latter area is frequently known as IRA 'bandit country'.

Some idea of the intermarriage totals for the South of Ireland can be obtained using the *Vatican Statistical Yearbook*, where figures for mixed marriages within a Roman Catholic church are available up to the mid-1970s. This publication provides information for the whole of Ireland and was also used by Lee for 1969–71 data. As the *Yearbook* gives 876 mixed marriages for 1969, 1,087 for 1970, and 1,382 for 1971, it does mean that for the South there may well have been about 500 mixed marriages or more celebrated in a Roman catholic church in 1968–9, over 600 in 1969–70, and over 800 in 1970–1.[2] Most mixed marriages in the South prior to 1970, the

[2] While undertaking research on the clergy's attitudes to intermarriage in 1973–4, I came to the conclusion on the basis of a rough estimate, using informally

date when new church legislation relaxing somewhat the restrictions was introduced, took place in Dublin. There was considerable local as well as clerical opposition in other parts of the country to the practice.

For the period of the 1970s the number of mixed-married couples living in the Republic of Ireland is likely to be greater than those marrying there. Not only were couples known to go to England to marry, but the South experienced net immigration from the UK for the first time in its recorded demographic history. Many of these were Irish people who had married in the UK, some to English husbands and wives.

It is perhaps important to point out the finding that Roman catholic women were twice more likely than the men to marry 'out' (Lee 1981: 74, 86; Compton and Coward 1989: 185–6), a pattern which has been noticed elsewhere, in Holland, Germany, and Switzerland. On the basis of a further analysis of Richard Rose's (1971) data from the famous loyalty survey, Lee suggests this may be largely attributable to protestant women's fears of marrying down the social ladder. But, clearly, part of the reason would be that the protestant women would have to face the likely loss to their church or ethnic group of their children in a regular marriage to a Roman catholic under catholic regulations then current in Ireland. The cultural fact is that in Ireland, North or South, the mother is primarily responsible for the religious determination of the children rather than the father, and this matter alone could have quite decisive impact. Group survival is clearly an important factor discouraging mixed marriages in Southern Ireland. Many Southern members of the Church of Ireland make great efforts to marry within their numerically residual church or to other protestants.

One might expect that mixed marriages were not very numerous in the three centuries from 1600 to 1900. But a number at least did take place. Lee is in line with popular historical opinion when he suggests that they were not uncommon: 'Giving evidence before a Select Committee on the State of Ireland in 1825, the Reverend Thomas Costello, a Catholic priest of Co. Limerick, observed that mixed marriages "were very frequent even among the lower

communicated data received from some 50 Roman catholic priests of the Dublin Archdiocese, that some 3% of all marriages in the area in 1960 were mixed and that this had risen to about 6% by 1970; that is from 200 to 400. This estimate was clearly too conservative.

orders"' (1981: 99). Even though such marriages were invalid by state law—protestants could only be married by a Church of Ireland clergyman—priests appear to have frequently performed such marriages. Comparative figures for the late nineteenth century can be obtained from Census records, where they still exist, as religious belonging figured in the counts by that time.[3]

It has been suggested by Irish commentators that the specific Roman catholic enforcement in Ireland of a papal decree on mixed marriages of 1908, restricting the possibility of mixed marriage quite severely and known as *Ne Temere*, had a dramatic impact on the size of the Southern Irish protestant community. Dr Garret FitzGerald, when foreign minister of the Republic in the mid-1970s, held the view that the Roman catholic legislation on mixed marriage did have quite a substantial effect on the declining size of the protestant population in Southern Ireland:

The effect of this very unbalanced ratio of religions in the population of the Republic has been that, despite considerable social pressures in favour of confining inter-marriage within the Protestant group of Churches, about a quarter of Protestants—at least in the 1946 to 1961 period for which a reasonably full analysis of data is available—have married Roman Catholics, and in the vast majority of cases during that period at any rate the children of these marriages were baptised into the Roman Catholic Church and brought up as Roman Catholics. The problem is complicated by the fact that the proportion of Protestant men contracting mixed marriages is higher than in the case of Protestant women and where the mother is Roman Catholic the likelihood of the children being brought up in that faith is especially high. The inevitable statistical or demographic consequences of this—in conjunction with a higher age level and consequently higher death rate amongst the Protestant population, although this is a relatively minor factor—is an erosion of the Protestant population by about 25% per generation—or around 1% per annum. No similar phenomenon appears to exist, at any rate on such a scale, anywhere else in the world. (1975: 189)

Part of Fitzgerald's argument is based on Brendan Walsh's statistical analysis of Census and other data for the populations of Ireland as a whole. Walsh concluded in 1970:

One reason for the extremely low O[ther] D[enominations] birth rate in R[epublic of] I[reland] appears to be the prevalence of mixed marriages in

[3] I am grateful to Ray Lee for this information.

which children with one OD parent are raised RC. It was estimated that about 30 per cent of all OD grooms and 20 per cent of OD brides married Catholics in Catholic ceremonies in RI in 1961.

The overall effect of these differentials in demographic behaviour has been to maintain approximate stability in the religious composition of the N[orthern] I[reland] population in recent decades, while the OD proportion of the RI population has fallen steadily. (1970: 33)

Walsh has since reappraised this analysis of the statistical data and now supports the corrected view that Roman catholic pressure on raising the children of such marriages as Roman catholics has had only a marginal demographic impact, if any at all (1975). Walsh's revised conclusions have been well received by the Roman catholic hierarchy. There is no doubt that it has let the church off the hook in what had become a rather embarrassing issue. Even so, the very data Walsh produced in the original argument still seem quite convincing.

The feelings of the Northern Ireland protestant–loyalist population on intermarriage are very powerful indeed. There can be little doubt that the popular fear of being outbred by catholic nationalists is still quite strong. This fear has been backed by research showing the higher rate of fertility of Roman catholics, though this rate is coming down. The significant dominance of protestant loyalists in the population has been maintained by high rates of emigration for Roman catholics, and this factor seems likely to continue to be significant. However, Compton and Coward do point out that while mixed marriages are very small in number, they do have some impact on the population structure, for half the children of such marriages have so far been brought up as Roman catholics and only a third as protestants.

Undertaking an intermarriage in Northern Ireland can have quite deadly consequences:

Dillon and Lehane . . . have documented various cases of assassination where the victims were involved in interreligious liaisons . . . There is even some evidence that social control may be mobilised quite effectively where an interreligious courtship has breached a territorial boundary. John Burns . . . in a journalistic interview with two teenage gang members in Belfast raised the question of 'mixed' courtship and in both replies, interestingly, the notion of 'going over'—'if she goes over to see him the other girls beat her up'—and 'getting over'—'we would wait for him and beat him up and make sure he doesn't get over again'—occurs. What is by no means clear, though, is whether or not such forms of social control are actively mobilized in less troubled times. (Lee 1981: 145)

Since Dillon and Lehane's documentation, such assassinations have continued. The kinds of problems experienced by intermarrying couples may be sometimes quite substantial in terms of social disapproval, even when physical violence is absent (McFarlane 1979).

THE CHURCH IN THE SPHERE OF MARRIAGE UP TO 1928

From Marriage

As a problem in the relationships between Christians in Ireland today, mixed marriage has a high profile. Members of the reformed churches find Roman catholic church attitudes and law highly offensive, though presbyterian regulations have also been negative. The sense of antagonism has also grown because protestants believe the Southern protestant population has shrunk in size as a result of Roman catholic interference and because of the bitter taste left by such past events as the McCann case, outlined in the next section. The churches hold to two main positions on marriage. For the Church of Ireland, its influence is as much due to the traditional Anglican role in civil society as to a theology which marries grace to nature. But the natural reality of marriage is still viewed as part of the created order, and this precludes certain forms of interference by the church. Even more than the Anglican, the presbyterian tradition stresses the natural reality of marriage and that the church has little role to play in it. Rather differently, the Roman catholic church has for several centuries tended to see church mediation of grace almost taking over from the natural marriage processes and this view sharpened rather than diminished in the first half of the twentieth century. These theologies may appear abstract in themselves but are quite real in their implications, especially when a Roman catholic wishes to marry a protestant. The couples find themselves on the receiving end of church discipline. The tensions and passions aroused indicate the close relationship established in the clerical and popular mind between theological and ethnic opposition.

Here a brief account of the history of the Roman catholic church's intervention in marriage and mixed marriage is revealing.

In the ninth century, the Western church had a different concept of marriage to the Eastern church. While the Eastern church already had strong control over marriage—a consequence of considering priests to be ministers of the sacrament of matrimony—the Western church saw the partners in the marriage as the ministers and their mutual consent as the basis.[4] The consequence of this position was that the Western church accepted as married any couple who declared themselves so, irrespective of public ceremony, profane or sacred. The judgement was based on the acceptance that marriage was primarily the partners' affair, though it eventually became illegal when not publicly solemnized. The period from the eleventh to the thirteenth centuries appears to have been one where the control of the church over sexual unions steadily increased, because of the role the church came to play in administering status, legality, and inheritance rights in medieval society (Duby 1978; more esp. 1984). But the lack of a need for witnesses caused very severe problems to this role. The frequency of clandestine marriages meant that legitimacy and inheritance rights were often in dispute and abandoned wives could not prove their status.

The problem remained unsolved for centuries until the Council of Trent in 1563. Because this council was convened specifically for anti-protestant action, coming as the Roman response to the Reformation, the legislation on marriage was to have anti-protestant implications. Marriage was declared one of the seven sacraments in response to the reformers' limiting of them to baptism and the eucharist, and there was a general stress laid on the authority of the church, particularly of Rome. This even greater sense of the church's mediation of divine grace affected its attitude to marriage. Whatever appeared to be natural about marriage was taken up almost imperceptibly into church law and the church was seen to have much more of a right than before in a marriage's material realities. The problem of clandestine marriages was dealt with in the decree *Tametsi*, by establishing a canonical form for all marriages, without which marriage in which at least one partner was a Roman catholic would be deemed invalid and not just illicit.

---

[4] See Pope Nicholas I's letter *Responsio ad Bulgaros* of 866: 'Sufficit secundum leges solus eorum consensu, de quorum coniunctionibus agitur [To fulfil the norms for a recognized marriage, the consent of the marrying partners is sufficient]' (Denzinger 1973).

The form required the exchange of marriage vows before a duly appointed cleric and a minimum of two other witnesses.

. . . to *intermarriage*

Though the Tridentine legislation was enacted to end clandestine marriages, the consequences for marriages between a Roman catholic and a protestant were already implicit. Such a ceremony would have to take place within the ambit of the Roman catholic church, otherwise there would be no valid marriage for that church. That Rome was likely to retain the legislation to bar intermarriage is implied by its already established attitude to marriages of church members with heretics, prior to the Reformation. When faced with mixed marriages in the period of the Roman empire, the church had insisted on the children being brought up in the church, and had tried to avoid such marriages ever taking place (Heron 1975: 100 n. 8). In the sixteenth century, the Roman catholic church had every intention of treating the reformed churches as, at best, heretical and, at worst, children of perdition. So, though the decree *Tametsi* had in itself nothing to do with mixed marriages, its provisions were to be interpreted by Rome as well in keeping with the way heretics had for a long time been treated by the church.

But the most important part of the decree *Tametsi* was that it explicitly assumed control over such marriage ever validly taking place, denying that there could be a natural marriage without people conforming to church demands. If one failed to observe the form the Roman catholic church explicitly laid down for its fulfilment, then one was deemed to be unmarried, even at the natural level and not just in terms of ecclesiastical discipline. This was, arguably, a power which the church had not believed itself capable of exercising before, and widened the gap between catholic and reformer. As will be seen below, the juridical understanding the Roman church had developed of itself over the years implied that legal dispensations from these conditions could be a possible way out in embarrassing circumstances. But the matter was to become even more complicated. The Council of Trent had stipulated that for the decree to apply it had to be published or promulgated in a given parish. It would be interesting to find out why such a stipulation was made. Whatever the results such an enquiry might show, the fact remains that many bishops and parish priests did not publish the decree for years, sometimes for centuries. As might be expected, this was particularly true of countries where Roman catholics were not overwhelming in numbers and where mixed marriages were likely to occur. One has here a prototypical case of the contrast between theological, legal rhetoric and concrete social ethics. Laws formulated at the political centre of the church appear to have had little bearing on the real circumstances of Roman catholics at the periphery, living cheek by jowl with heretics. It is not surprising then that in 1741 some areas such as the Netherlands and Belgium became explicitly exempted by Rome from the canonical form prescribed by Trent in so far as inter-church or 'mixed' marriages were concerned (see Benedict XIV, *Benedictina*, in Denzinger and Schonmetzer 1973). Disputes and differing points of view between moral theologians on validity and legality did occur (Connick 1964). Though canon lawyers do not mention it, it is highly noticeable to a sociologist that the moral theologians and canonists who took time to examine the moral and validity issues of mixed marriage were, in the main, Spaniards: and the King of Spain was the ruler of the Netherlands throughout the Reformation and early post-Reformation period, when and where such matters were live issues. Whatever the opinions, Rome and its moral theologians did seek to enforce a general backdrop to decision-making whenever approached, which seems to have been rarely. If a mixed marriage was mooted without the conversion or promised conversion of the non-Roman catholic partner, then it was only to be allowed to take place so long as the danger of perversion to the faith of the catholic party was removed. There had to be an undertaking to bring up the children of the marriage as Roman catholics and a dispensation forthcoming from Rome. Otherwise, any such attempted marriage would not only be immoral and illegal according to church law, but invalid as well. In the seventeenth and eighteenth centuries, Rome only granted dispensations if the heretic actually converted. In other words, the baseline from the centre was the absoluteness of the centralized and authoritative judgement of the Roman catholic church and the perverseness of all alternatives.

In reality, however, local bishops also had considerable power, and frequently appear to have permitted such marriages even without the conversion of the protestant partner. Before Clement XI's intervention in 1710, refusing a dispensation in a particularly benign case (Connick 1960: 309 ff.), and setting the pattern for any future requests, many canonists apparently removed from the geo-political centre had made the distinction between cessation of ecclesiastical law *ab intrinseco* and cessation ab extrinseco, the former due to Roman dispensation, the latter to local practice or common custom. But from the late seventeenth century, the majority of canonists came to consider such custom-approved marriages as both illegal and immoral, even if valid, whatever bishops might permit. This became known as the position of canonical tolerance. By now the Roman line was crystal clear: Cardinal Petra (1662–1747) argued that, even for those holding such marriages as licit and moral, the promise to bring up the children as Roman catholics had always been necessary, though whether Petra was correct on this matter is a moot point. This Roman position was to remain all the way up to the second Vatican Council (1962–5), and in some countries even after that date. But local bishops and clergy were still known to dissent in peripheral areas. For example, in Germany bishops did not publish the *Tametsi* decree and dealt with matters themselves, not requiring conversions. On the side of the liberals was the fact that Rome had granted dispensations in the past and recent present for royal marriages, so the practice could hardly have been immoral of itself. It is clear that the publishing of the *Tametsi* decree would have made things considerably more inflexible. For if the form for such a marriage was not adhered to, what was a matter of morality and legality—being right or wrong—then became a matter of validity, and such a marriage was not considered a marriage at all. Here, the very passage from legality to validity is an index of growth of the sense of power the centralized church felt it had over the sexual and family lives of its faithful. And yet, if Rome was able to quash the application of the decree in certain areas where it had already been published, which it did in some cases, was this not in fact an implicit recognition that there were limitations to its power over marriage as such and not simply to its ability to coerce? Canonists argued that it was simply an application of the 'principle of double effect': allowing the evil of mixed marriages while at the same time avoiding the greater evils of civil unrest, clashes with civil authority, and a worsening of the conditions of local catholics under protestant rulers. There is always a logic to the monopoly position of thought.

Summing up, one can say that, up to the end of the eighteenth century, while it appears that there was an attempt to set up a general opposition to mixed marriages from the centre of Roman catholicism, there was also an attempt to weaken this approach at the periphery, one which Rome was often forced to accept.

In Ireland the decree *Tametsi* was promulgated in the four ecclesiastical provinces at different times: in the provinces most affected by the plantation, those of Armagh and Tuam in the North and North-West of Ireland, the decree was published in the sixteenth century. But in the other two provinces the decree was not published for 200 years: in Cashel in the South-West, in 1775, and in Dublin, in 1778 according to Lee (1981: 13) and in 1827 according to Corish (1985: 135). Also according to Corish, the decree only came into force in the wardenship of Galway and the diocese of Meath in 1827. But the effects of this publication and non-publication are apparently little known or little researched. Again, Roman catholics both clerical and lay considered mixed marriages before a Church of Ireland minister as valid, not least because the penal laws rendered a mixed marriage civilly invalid unless thus celebrated. In 1785, Pius VI declared mixed marriages in Ireland to be exempt from the form for all other marriages laid down by the Trent decree. Clearly, part of the reason was the avoidance of conflict with Irish civil legislation which did not recognize catholic–protestant marriages performed by a Roman catholic priest. But there is no absolute certainty that this was the only reason. There could also have been pressure brought to bear from local experience itself over the moral questions at issue, or, more plausibly, the civil problem may have provided the excuse to embrace wider local practices.

S. J. Connolly affirms that there was 'a clear hostility of the church authorities to intermarriage' (1982: 196–7). But among the upper classes at least, there was a customary agreement that a boy followed his father's, and a girl her mother's, religion.[5] Edwards (1970) suggests the practice was quite widespread and that it was probably related to that system of marriage whereby men rather than women marry down the social scale. In any case, the custom of boys following the men, and girls the women, would have ensured that rights of inheritance remained within the family or origin (Lee 1981: 102). Given the importance of the structures of inheritance for the economic interests of the family, it would seem quite

---

[5] Corish 1984: 220 appears to include the lower classes in this custom.

plausible that the contradiction between these and the orthodoxy required by Rome might have resulted in contradictory practices.

It is also clear that priests did supervise mixed marriages despite the civil invalidity and despite thus committing a felony.[6] As one priest declared, catholics were 'quite unwilling to legalise their marriages with Protestants, by having the rite performed in the Protestant church' (Parliamentary Papers 1825: 425, quoted Lee 1985: 13).[7] The custom notwithstanding, Irish bishops appear to have taken different views of mixed marriages. In 1825, when giving evidence to the Select Committee on the State of Ireland, Bishop Doyle, the well-known nineteenth-century Roman catholic ecumenist, affirmed that his church imposed no obligations on such couples, while Archbishop Murray of Dublin affirmed the contrary. Murray's view—and Murray was very much for an Irish line more independent from Rome—can be considered supported by the testimony of a Church of Ireland clergyman, Revd Mortimer O'Sullivan, who furthermore added that the protestant partner often yielded to such demands, particularly where the bride to be was the Roman catholic (Lee 1981: 101).

In any event, the differences between Doyle and Murray were testimony to some form of divergent practice, possibly along the lines of a North–South split, one which had already resulted in the separate introductions of the Tridentine decree, as well as to the high degree of autonomy the Irish bishops up to the Cullen era appeared to have exercised both in respect of Rome and of each other.

The Synod of Thurles in 1850 appears to have resolved the issue for the clergy and bishops: it 'recalled' the obligation on priests to extract the promises from the partners before permitting an intermarriage (Lee 1981: 102; Connolly 1985: 27) and made the disapproval of such marriages felt by rendering the liturgy for such occasions very sparse and empty of ceremony. The change was largely due to the influence of Paul Cullen, then Archbishop of Armagh. Cullen even got the other bishops to accept at Thurles the need for a papal dispensation. But when, three years later, the Westminster synod in England received permission from Rome to

---

[6] The offence was only withdrawn from the statute book in 1833.
[7] One should remember also that marriages between presbyterians and Roman catholics too had to be performed by a Church of Ireland minister up until 1844 according to civil law. Cf. Corish 1985: 124, 220.

leave dispensations in the hands of local priests, it also became business as usual in Ireland as Irish bishops proceeded to obtain the faculty for themselves. What is also very important to stress is that, even if such a marriage took place without the requisite promises, such a marriage was still recognized as valid by the church. The practice of a son following his father's religion and a daughter her mother's may have continued for some time even after the Thurles decision, though there is no published evidence either way, to the writer's knowledge. One gets some taste of the state of clerical attitudes after the middle of the nineteenth century had passed, from Bishop Moriarty's statement to the Royal Commission on the Laws of Marriage in 1866:

if I were to consult my own feelings and convenience I should not wish for any change in the [civil] law [forbidding mixed marriages to be celebrated other than by a Church of Ireland clergyman] because it is very convenient for me, wishing as I do to prevent such marriages to be able to say to parties seeking our ministration, 'No, the marriage will not be valid in law and I shall be subjected to a severe penalty' ... (quoted Lee 1981: 100)[8]

Civil legislation was amended in 1870, rendering the ministry of Roman catholic priests legal for mixed marriages.

Hence, during the nineteenth century the central Roman theology on mixed marriages began to affect the Irish scene. The consequence of the Council of Trent's decree was that what had previously been considered a potentially dangerous undertaking for the faith of the Christian partner and for his or her children now became an undertaking which the church felt it had the power to prevent being a marriage at all. The step could be quite momentous in the civil and political sphere, for it was now possible for a marriage to be declared valid by the state and invalid by the church. Such an eventuality had been largely avoided by the Roman church occasionally receding from the brink and by the deference to local custom, so that there was a degree of rhetoric about this power over validity.

But all this changed in 1907 when the decree *Ne temere* was issued from Rome. This new decree reversed all the exceptions to the original *Tametsi* decree. It made the Tridentine decree universal in that, on its publication in an ecclesiastical province or by a duly

---

[8] The quote probably gives the flavour of the period, though Moriarty was also one of Cullen's men.

constituted hierarchy, it made all marriages involving a Roman catholic dependent for their validity on the officiation of a priest and the presence of at least two witnesses. Also, in the case of a mixed marriage, there had to be granted previously a dispensation from mixed religion and the granting of this dispensation depended on certain promises being made. The actual practice was to be as follows. When the priest was first approached to agree to marry the couple, they were to be asked if they would promise to baptize and bring up the children in the Roman catholic church. If the couple agreed, then the priest would apply for the dispensation from mixed religion. The promises thus in practice became a prerequisite for validity itself. Also, from now on, the cases of the intermarried were explicitly included in the general legislation rather than otherwise so that no local exclusions were possible.[9] Canonists have argued that the introduction of the new regulations was merely a 'tidying-up' exercise of canonical laws. But of course canonical housekeeping implies a genuine extension of power. In reality, therefore, the Roman catholic church was now extending its power over marriages, and to the detriment of inter-church relations.

Still one further change was included in the new rules. This was the most significant, as the Roman catholic church did something it had never attempted before, namely render invalid certain marriages which had taken place before the degree was promulgated. Previously contracted, illegal marriages between a catholic and a protestant were made invalid by retroactive legislation: and one cannot really have a greater sense of power in such matters than that. This legislation in particular became very important in Ireland with the McCann case, which occurred barely three years later in Ulster.

THE MCCANN CASE AND INTERMARRIAGE UP TO THE SECOND VATICAN COUNCIL

The main currently published source and only sociological evaluation of the McCann case so far is Lee 1985. Lee tends to see the furore which surrounded the case as unjustified and an expansion

---

[9] *Ne Temere* explicitly excluded Germany and Hungary from its effects, but when the *Code of Canon Law* was brought out in 1917, these were also included; though marriages between Roman Catholics and eastern Christians were exempt.

of the fears of Irish independence with which protestant loyalists were faced at that time. The present writer differs on this point. The case can be seen in a critical perspective, namely as the result of the concrete violation of human rights which enforcement of Roman catholic church legislation entailed. It is partly for this reason that the case had an immediate consequence of arguable proportions in consolidating the protestant–loyalist alliance. Lee gives a brief summary of the event.

The public first became aware of the McCann case in November 1910 when a letter from the Rev. William Corkey, Minister of Townsend Street Presbyterian Church in Belfast, was published in the local press . . . Mr. Corkey's letter concerned the plight of Mrs. Agnes McCann, a member of his congregation. Agnes McCann, the letter stated, had married her husband Alexander, a Roman Catholic, in a Presbyterian ceremony some years before *Ne Temere* came into effect. The couple agreed at the time of their wedding that they should attend their respective churches, and the marriage itself was a happy one. Eventually, however, the letter went on, Mr. McCann's priest came to the house and told the couple that the marriage was invalid as a result of the change in canon law brought about by *Ne Temere*. Mr. McCann asked his wife to remarry in a Catholic ceremony and when she refused, began to ill-treat her. Some time later, he deserted her, taking with him the couple's two children and, later, all of the furniture. The priest, who regarded Mrs. McCann as having been living in sin, would do nothing to help her, leaving her destitute and without her children. (1985: 16)

On publication of Corkey's letter, there was an immediate reaction from Northern presbyterians who held special meetings, including one of the Belfast presbytery, to express their horror at the event and to organize and seek redress for the hapless victim. There are really no other published sources of the facts other than those provided by the press and Corkey himself. Mrs McCann died a few years ago and it is unlikely that there are other informants still surviving. Lee argues that the event was amplified, perhaps even unwittingly, in order to show up the Roman catholic church in the worst possible light: 'What had happened to Mrs McCann was almost invariably interpreted on the basis of an assertion that the Catholic Church was an aggressively domineering institution' (1985: 19). We have seen in the previous section that there clearly was something domineering about the enactment of retrospective legislation in respect of existing marriages. There can be little doubt that the local priest did what he did and that Mrs McCann ended up being abandoned (cf. Barkley 1972: 97). These facts were not in dispute. Whether the request from the priest to be remarried shook the foundations of the marriage or whether it was a useful excuse for Mr McCann to leave on whatever grounds, we shall never know. Irish Roman catholic MPs argued in the House of Commons that Mrs McCann had in fact persecuted Mr McCann, for example, by having meat ready to eat on Fridays instead of fish (Lee 1985: 22). It is true that the priest who visited Mrs McCann was never produced, and she refused to name him. However, the procedure on the Roman catholic side is predictable according to the logic of the legislation, and there can be no doubt that some attempts to get mixed-married couples to remarry were made. Also, the very statement that mixed marriages were invalid, offended presbyterians on a number of counts. The first, mentioned by both Lee and Barkley, is that presbyterians had only been allowed by civil law to celebrate mixed marriages since 1845, and hence a particularly offensive twist could be seen in the fact that now the Roman catholic church, as opposed to the Church of Ireland, was referring to some of their marriages as invalid. But, secondly, there was also the fact that for presbyterians marriage was a natural institution, and any church's right to interfere with it was severely restricted. The Roman catholic church's attempt to declare a marriage invalid, particularly retroactively, was therefore doubly insulting, not only to them but to the will of the Creator. This on its own would have been enough to cause a religious furore.

Lee is, however, right to see the protestant view of the event as precisely the kind which would harness it to the global interpretations of political religion among the protestant loyalists as a whole. Barkley (1972) could be right in saying that the entry of the presbyterian church leadership into the loyalist camp was prompted by the McCann affair. It was only after that the general assembly, where debates on the political situation often included references to the McCann case, swung in favour of opposition to home rule (Barkley 1972: 97, 238), eventually voting overwhelmingly against it in 1913. But perhaps one should also add that the whole movement of polarization was taking place at this time, with the Liberal Party in Britain yet again bringing forward legislation to enable home rule and protestant loyalists responding with their solemn league and covenant in 1911, which was still being signed in presbyterian halls in 1912 (Barkley 1972: 97, 238). Corkey, and those who shared platforms with him in the protest meetings on the McCann case, also accused the Roman catholic church of trying to undermine the law of the state—British law to boot—thus offending the other foundation of the protestant–loyalist dominant ideology at the time: British imperialism and the British cultural heritage as a whole. The threat was also seen as going to the very basis of natural society itself, the family. There is, of course, logic in the protestant–loyalist argument at this juncture. Whatever the source of protestant–loyalist ideological armaments, this bullet at least was real and live. The protestants of Ireland were aware of the political power being brought to bear by such legislation and of the human rights thus being violated, no matter how few people it affected in reality.

But their Roman catholic counterparts, probably in all truth, could not see this violation, as for them the material and social truth implied in the move was hidden from them by arguments on the tidying up of canon law which *Ne Temere* implied and on the rights and divinely granted power of the church to fix the form of the sacrament of marriage and to ensure that natural and divine law was observed when it came to educating the children of such marriages.

The *Ne Temere* decree was in force all the way up to the Vatican Council of 1962–5. In Ireland, it was rigorously adhered to in both spirit and letter. Mixed marriages were probably extremely rare, except where the protestant party was prepared to convert, a principle which apparently was drummed home into the hearts and minds of Irish catholic youth in the intervening years. Mixed marriages outside of Dublin were largely unheard of, and bishops outside the city generally never gave permission for one. When mixed marriages did take place, and the protestant partner refused to convert, the extraction of the promises was the prime object without which the marriage was not approved. The promises had to be made in writing and by both parties. Marriage only in a catholic church, or rather in a side room of the church, and before a priest was permitted. The idea was that there was to be no solemnity, no singing, and no celebration. Disapproval had to be felt and not just announced. In terms of protestant rights, there was simply no awareness that an adult who has no children as yet cannot understand the bonding, care, and consequent desires to share values and beliefs with them, and that making promises for the future of their children is thus radically flawed at the moral level. An indication of the deeply ingrained Southern consensus on such issues was the Tilson case of 1951–2. A protestant, Ernest Tilson, had married a Roman catholic and taken the written promises, agreeing to his wife bringing up the children of the marriage as Roman catholics. Nine years into the marriage, he changed his mind and sent the three children to a protestant school. The wife took the family disagreement to court, where Judge Gavan Duffy found against the husband on the basis of the promises and despite common law favouring the husband as head of the household. On appeal, the high court confirmed the verdict, with the one protestant judge among the five dissenting from the judgment. Combined with the ruling on adoption seen in Chapter 5, the verdict also strengthened the practice of catholic adoption agencies prohibiting mixed married couples from adopting any of their children.

That is the way it was, and to some extent remained in the Republic of Ireland even after the second Vatican Council. Clearly it would be insufficient to see the rigid ethnic divisions as the product of one church's legislation. Rather it should be seen as an element in a range of religious–political relationships of antagonism, but one which tidies up the boundaries of separation. It provided a thoroughly religious definition of the taboo of crossing the boundary via sexual liaison. The legislation of *Ne temere* was adding another important dimension to the already existing state of alienation between the two groups, adding further to the religious significance of the conflict.

THE SECOND VATICAN COUNCIL: THE LOOSENING OF RELIGIOUS MONOPOLY

Religious liberty and the recognition of other churches were central themes in the deliberations and conclusions of the second Vatican Council. They affected not only the Council's final documents on *Religious Liberty* and *Ecumenism*, but also the dogmatic constitution on *The Church* and the Council fathers' decision to reform mixed marriage legislation in November 1964. The best way to take stock of these developments within the minds of Roman catholic prelates is to look at the dominant, competing, and subordinate ideologies of the church as they were at that period of time. The canonical view and that of the first Vatican Council of 1870, which declared the infallibility and primacy of the Pope, was thoroughly monopolistic and Rome-centred. It was also most effective in the sphere of marriage and mixed-marriage regulations as found in the Code of Canon Law which made *Ne Temere* universal to the latin rite in 1917. That dominant ecclesiology can best be described as a set of variations on one central theme, namely that the church is essentially defined by its hierarchical authority structure with its most important member being the Pope at the apex and the varying ranks of clerics subordinated to him and arranged in a downward chain of command.[10] An additional theme of the period was that of the church being one of two powers ordained for the control and administration of human society, the spiritual as opposed to the temporal power—very much a theme retaining a certain continuity with the medieval conception. But this view of the church was enhanced by the interaction of doctrinal development with the historical crises of the Vatican of the nineteenth century, particularly the breakdown of papal temporal power. Consequent on the unification of Italy and the proclamation of the doctrines of papal primacy and infallibility, there resulted the clear tendency to consolidate and centralize more effectively the remaining power of the papacy throughout those areas of the world where the Roman catholic church still held sway. The official reason for the standardization and systematization of the church's statutes into the Code of Canon Law, promulgated as such for the first time in 1917, was that the number of regulations governing the universal church and its various parts had grown too large and had become unmanageable. But one suspects that motivations of power and control underpinned such ordering. In any case, intellectual ordering by organic intellectuals is always related to issues of power. We have already seen how this vision was applied to mixed marriage and how the church conceived of itself so powerful as to render mixed marriages invalid. As Heron says, 'There is of course no parallel to these provisions in the law or practice of the Protestant churches...' (Heron 1975: 84).

What became apparent in the debates and documents of the second Vatican Council was that the conception of a church which is wholly centralized and seeks to standardize rules almost

---

[10] For an example of a theological textbook of the period couched precisely in these unquestioned terms, see Van Noort 1960.

irrespective of cultural variance was no longer acceptable to the majority of the council fathers. Many of them were in the business of reshaping the national church to local issues and in strengthening their local control to the detriment of Roman domination. Hence the old canonical and centralized view no longer appeared as the only possible theology of the church which Roman catholics could adopt and other theologies were found to be acceptable. Of course, if applied these would have a different effect on mixed marriages. One principal element in these lately grown ecclesiologies hinged on the perception of whether the true church was the Roman catholic church or merely 'subsisted' in the Roman catholic church. The council fathers, in the dogmatic constitution *The Church*, were careful not to identify 'the unique Church of Christ' with 'the Catholic church' or 'the Holy, catholic, and apostolic Roman Church'. In fact, the first draft of the constitution presented by the appropriate commission and which proclaimed the identity, was rejected by the council, which substituted instead the word 'subsists' for 'is':

Christ . . . established and ceaselessly sustains here on earth His holy Church . . . This Church, constituted and organized in the world as a society, subsists in the Catholic Church, which is governed by the successor of Peter and by the bishops in union with that successor, although many elements of sanctification and of truth can be found outside of her visible structure . . . (Vatican II 1967: 136)

The council itself gave no exegesis of the term 'subsists' and so the implication of such a position was clear: a wide range of theological positions could be legitimately formulated, as Coventry (1975) has indicated. Of course, if any of these alternative ecclesiologies were to be embraced by the papal centre, or if several were allowed to compete for political dominance, cosmo-political catholicism, that is one which seeks universal conformity, would be finished. It is of course not the case as yet, and there is abundant evidence that there is no intention of the cosmo-political centre abandoning its most cherished role in the world at large, while at the same time drawing back from total implementation of the cosmo-political ideology, for example in the face of liberation theology in Latin America. Such developments point to a political struggle between factions in the Roman catholic church as well as to a struggle by the periphery for more control. Thus the debates were not non-political or purely theoretical.

Whatever the outcome of this macro struggle for power, the problems for individuals which rigid Vatican control of mixed marriages had brought were clear enough. Certain directives were to become important in their regard. These were the documents on ecumenism and religious freedom. These recognized that the rights of the Roman catholic church could no longer be taken as absolute in respect of those of either individuals or other religious groups. It was in this spirit that a number of remarks were made by council fathers on mixed marriage and that the Roman secretariat and the Pope were directed to draw up new rules governing mixed marriage. But those who still tended to stress a hierarchical framework for their ecclesiology could still find some support in affirmations of the dogmatic constitution on the church, as pieces referring to its hierarchical constitution were inserted at the behest of Paul VI into the final draft, with the council fathers' agreement. Hence, those who tended to a conservative stance on the mixed-marriage issue had something to fall back upon. Those who saw such marriages in a new light could call on the priority given by the document to the people who make up the church, both good and bad, as well as to the importance of the rights of other churches and their baptized members.

Mixed marriages between two Christians steadily became known officially as 'inter-church marriage'. It was also more noticeable that such marriages between deeply committed Christians tended to encounter more resistance from the Roman catholic church than marriages between indifferent Christians. This came to be seen as nonsensical; the difficulty was not the fault of the couple, but of the churches themselves in so far as they were divided.

In answer to the council fathers' request to reform the mixed-marriage legislation, the Vatican published in 1966 a first document[11] on the subject, whose guidelines were to become incorporated in the new Code of Canon Law of 1983. The instruction tended to repeat the traditional teaching that mixed marriages were to be avoided wherever possible. A dispensation was still required. Safeguarding the faith of the Roman catholic partner was a prerequisite, and the future Roman catholic upbringing of the children was to be guaranteed. The local bishop and priest were to be responsible for educating the Roman catholic partner in the

---

[11] Vatican II 1967: 1130–9. The document is from the Sacred Congregation for the Doctrine of the Faith, 1966.

responsibilities of catholic baptism and education for the children, and for obtaining an explicit promise to this effect from the Roman catholic partner. The non-Roman catholic partner had to be informed of the Roman catholic teaching on marriage and mixed marriage, and still had to promise, in writing, not to interfere with the catholic upbringing of the children. However, there were some clear relaxation of discipline which amounted to an increased awareness of non-church members. If the non-Roman catholic felt in conscience unable to promise, then the matter was to be referred to Rome, implying that such a promise could be waived. Also, although the promises were expected to be made in writing, the bishop could determine otherwise case by case, and also decide whether the promises should appear in the marriage documents themselves. It was also allowed for the children not to receive a catholic education in the cases where a contrary local custom existed, and for the local bishop to dispense from his demand 'so long as the Catholic party is willing, in so far as he/she know how, to do everything possible' to fulfil their responsibilities (Vatican II 1967: 1136–7). The last point was especially directed towards the situation in eastern Europe. The canonical form was to be observed for the validity of the marriage, but if there were any difficulties in this respect, the matter was to be referred to Rome, implying that a marriage could take place in a non-Roman catholic church and before a minister other than a priest. Such marriages could also now be celebrated in church, with the church's rites, blessing, and sermon. After the service the 'non-catholic' minister could speak and prayers could then be led together. The responsibility of due pastoral care of the marriage was urged on local priest and bishop.

The instruction has been detailed because, though not a volte-face, it did represent significant changes. Despite the rhetoric of maintaining the original, traditional theoretical position, in place since the Council of Trent in the sixteenth century, the natural reality of marriage was creeping back in, in the newer formulation of human rights issues. It was now clear that any absolute and premeditated undertaking to bring up the children as Roman catholics was now considered too great a demand and probably unreasonable, if not unjust: but one rarely finds the Roman Catholic church admitting errors on its part.

But such was the opposition to the document from certain bishops who had sought at Vatican II greater liberality than that now proposed that by 1970 Pope Paul VI was forced to supervise a revised document, an apostolic letter entitled *Matrimonia Mixta*. The document gave over considerably more power to local bishops in the matter of mixed marriages on the grounds that discipline on such a matter 'cannot be uniform'. In addition, it was recognized that 'both husband and wife are bound by [the] responsibility [of educating their children in the same faith as themselves] and may by no means ignore it or any obligations connected with it' (Paul VI 1970: 795). The equality between spouses' consciences was thus recognized. This now had to be a motif behind any commitment on the part of the spouses. National episcopal conferences were to establish a common set of conditions under which they would grant dispensations. In addition, the phrase used to describe the Roman Catholic party's obligation to bring up the children as Roman catholics was changed from 'to do everything possible' to 'to do all in his/her power'. This was frequently understood to mean that any efforts on the part of the Roman catholic partner had to be within the context of the other party's parental rights and duties, as well as in terms of the growth of love and respect between the partners, a growth which was not to be harmed in any way. Thus the way was open for a variety of liberal interpretations. Bishops' conferences in Germany, France, and Switzerland set up joint working parties with other churches to produce agreed procedures (see Hurley 1975: 120–44). Despite the official disapproval of mixed marriages, at least one conference of bishops, the Swiss, appeared to dismiss this baseline as too negative and based their instructions on what they saw as the natural, positive fact of two people wishing to marry (see Hurley 1987: 127–34). Things at the cosmo-political centre were changing at least for mixed marriages. It was now becoming possible for the local Roman catholic church to deal with mixed marriages within the context of the local, social, and political milieux.

CLERICAL PRACTICE AND ATTITUDE IN IRELAND

In developing their local norms, the Swiss bishops submitted a first draft to representatives of the reformed churches for comment and subsequently modified them in the light of their remarks. They were also of the opinion that 'mixed marriage can also contribute to the restoration of Christian unity', and that the children were to be brought up in a way consistent with respect for the religious convictions of the non-Catholic partner and without prejudice to the marriage union. 'We note with joy that a common Christian faith makes a common Christian upbringing possible' (quoted Deane 1974: 546). The Swiss also allowed for the assistance of the pastor of the non-Roman catholic partner in the ceremony and for the Roman catholic priest to assist in the ceremony if it took place in the church of the other partner. Joint pastoral care was then provided for. The response of the Irish bishops in the same year consisted of a one-page document which made no similar concessions, but instead stressed that 'A Catholic can never be dispensed from the duty of handing on to his children the faith that is in him ... Lack of agreement about matters so deeply affecting family unity can constitute a danger of subsequent disharmony in married life' (Irish Episcopal Conference 1970). The bishops then went on 'to emphasise that, unless a Bishop's authorisation has been obtained, the marriage of a Catholic other than in the presence of a priest and two witnesses is null and void'. There is no denying the wholly negative tone of the communication and the insistence on the need to retain 'fidelity towards Catholic truth'. For several years, the Irish bishops were unable to agree to a common set of norms. It was widely known that one or two bishops would not countenance a mixed marriage at all in their dioceses, never mind agree to specific conditions. Notably, not only did the bishops require promises to bring up the children as Roman catholics, but these had to be made in writing and the protestant partner was required to sign that he at least acknowledged the Roman catholic partner's duties in this respect. When a set of norms eventually came in 1976, they were merely a subset of regulations within a wider *Directory on Ecumenism in Ireland*. The written promise was still required though not mentioned in the directory. What was mentioned was the obligation by divine law, which could never be dispensed, to do all in his or her power to baptize and bring up the children as Roman catholics. So strong did the bishops feel about this that, even in their later document of 1983, they talked about couples needing to think again about marrying at all, if they had not settled the matter of the children's upbringing beforehand. Despite this rigid interpretation, one or two Irish bishops were giving occasional dispensations to marry in the protestant partner's church. This fact suggests that part of the function of the directory was to keep certain ultra-conservative bishops happy. Clearly, another part of its function was to retain what appears to have been the most conservative approach to the problem in the Roman Catholic Western world.

The reaction of the two main protestant churches in Ireland was largely predictable. The presbyterians continued to operate according to the rules they had previously formulated. Their *Book of the Constitution and Government* affirmed: 'It is recommended that a minister do not join in marriage a member of the Church with a member of the Roman Catholic Church, nor with a person professing anti-Christian views, until in each case profession of adherence to the Presbyterian faith is made, after careful instruction has been given . . .' (para. 260, quoted Heron 1975: 100). Considerable resistance came from the Church of Ireland. Many ministers refused to assist at a wedding between members of the two churches if any promise by the Roman catholic about the upbringing of the children had been made, and they were supported in this by their otherwise ecumenical archbishop of Dublin.

Criticisms of the Roman catholic bishops also came from within their church, notably by the Jesuits Declan Deane and Brian Lennon (1980), and a Cistercian monk, Eoin de Bhaldraithe. It was pointed out that the continental solution to the mixed-marriage problem adopted in France, Germany, Benelux countries, and Switzerland, also permitted some of the children of a mixed marriage to belong to the non-Roman catholic partner's church. It was also said that an anti-protestant spirit still affected the Irish bishops' thinking on the issue.

Meanwhile, at the cosmo-political centre, another battle on the issue occurred. In 1981, at the biennial meeting of the synod of bishops, set up after Vatican II, considerable disquiet was expressed that the present directives on mixed marriage did not go far enough. It was agreed by the 200 bishops that a more open approach should be written into the new canon law, due out in 1983: more powers to local bishops' conferences; differentiation between the Roman catholic spouse's obligation to preserve personal faith and the obligation towards the children; stress on bearing witness to one's faith 'through the quality of love shown to the other spouse and the children; recognition of the positive ecumenical quality which such a marriage can have (de Bhaldraithe 1981: 642–3; *The Tablet*, 7 Feb. 1981: 142). But when the new

Code of Canon Law appeared, it tended to stress the document of Paul VI (1970) rather than the feeling of the Synod of 1981. Positively speaking, there was a greater concern with the pastoral aspects of these marriages, an open acknowledgement of the rights and duties of both parties, and the reaffirmation of the freer hand which local bishops' conferences had been given.

The Irish bishops responded with a fully worked out *Directory on Mixed Marriages* together with an accompanying handbook for those preparing for such a marriage. In some respects, the approach was old-fashioned and rooted in an absolute position. There was the stress on the difficulties of mixed marriage: 'It is a key element in the Catholic Church's pastoral approach to this matter to seek to avoid . . . a blurring of the issues by helping the couple to face, and seriously reflect upon [the] difficulties *before* the marriage. . .' (Irish Episcopal Conference 1983c: 1). But in other respects, the Irish bishops either took on or permitted the practice of putting the couple before the church, though they only did so in the pastoral handbook, which neatly side-stepped conservative rhetoric: 'As a general principle the Church discourages mixed marriages . . . You are not a "general principle", but a couple who at this stage share a great deal at the level of affection and mutual understanding' (Irish Episcopal Conference 1983b: 5). The directory also elaborated the distinction between the marriages of loosely committed Christians and those of deeply committed Christians. By so doing it admitted the more positive features of the second kind of marriage from a Christian point of view. However, a great deal of time was spent in reaffirming the obligation in conscience for the Roman catholic partner of doing all in his/her power for the baptizing and bringing up of all the children of the marriage as Roman catholics, and, equally, on deciding on this matter before the marriage. After several pages on these, a paragraph is added accepting that in practice, during the marriage, this may turn out not to be possible and that, in this case,

the obligation of the Catholic party to share the Catholic faith with the children does not cease; it still makes its demands. These would include playing an active part in contributing to the Christian atmosphere of the home; doing all that is possible by word and example to enable the other members of the family to appreciate the specific values of the Catholic tradition . . . (Irish Episcopal Conference 1983b: 21)

While the handbook is a more positive document, it is clear that the directory is aimed at curbing what are seen as enthusiastic ecumenists and in reasserting as far as is now possible the monopoly catholic position.

To some extent, the unquestioning blanket approval of endogenous ethnic marriage formerly implied in the practice of handling mixed marriages is now questioned within the Irish clerical church. Much of this change in approach can be seen to come from an inter-church working party the Irish churches had set up and which had reported in the terms the new handbook had sought to embody. In fact the practice is considerably more liberal than even the document suggests. In the majority of cases in the Republic, it would appear that the wedding takes place in the church of the bride, with her spouse's minister assisting in a number of cases. As a response to the easing of the Roman catholic church's position on the bringing up of the children, the negative attitudes of both the presbyterian and Church of Ireland leadership and clergy on the matter have largely disappeared.

Of course, in Northern Ireland the situation is much more difficult and the practice of spouses coming South to marry is likely to continue for the foreseeable future. What continues to be the problem throughout Ireland and in addition to the specific problems created by threats of violence and actual murders of the intermarried are the negative attitudes of kith and kin to such marriages. In other words, after years of religious support for the absolute divide between catholic and protestant, it is going to take a long time before the religion of the people modifies to allow the breaching of ethno-religious boundaries without sanction from either family or local community.

In the introduction to this section, the possibility of Roman catholic legislation and practice on mixed marriage having an effect on the size of the Republic's protestant population was raised. It does seem to this writer that two points have been seriously overlooked in the debate. The first point reiterates an earlier observation. In the Western world, there is no state other than Hungary where a significant minority of protestants exists alongside a majority of catholics. The second point is that the very legislation on mixed marriages in 1908 had a significant consequence in the public sphere and for the perceptions of mixed marrieds or intending mixed marrieds. For the first time, the

Roman catholic church in Ireland was declaring marriages which had already taken place as well as marriages in the future not to be marriages at all. The church was thus declining to accept the vows of those who had already taken them, along with all the moral responsibilities towards the spouse and children that these indicated. The retroactive nature of the legislation of 1908 was seen by protestants in Ireland quite rightly as a basic attack on civil society, marriage, and the family, especially in the McCann case. These dimensions of the debate in Ireland have hardly been appreciated on the catholic nationalist side at all, which tends to see the issue as a case of protestant exaggeration. In addition, no matter what other mores might prevent catholic nationalists marrying protestants in Ireland—the vision of the others being bigots, or oppressors, or ethnic aliens—the position of the Roman catholic church on the immorality of a mixed marriage for a Roman catholic, unless the promises were taken by the Roman catholic and assented to by the protestant partner, and the undesirability of the marriage in any case, constituted by themselves a divider between catholic and protestant and a promoter of their distinct and separate cultural processes. The more catholic a Roman catholic was, and the more he or she heeded the teaching of the church, the more likely it was that this separateness, combined as it was with the desired separate school system, would be maintained and promoted. Even now, with the degree of restriction remaining, there is sufficient displeasure in clerical attitudes to legitimate the negative elements in the culture of catholic nationalists.

8
Conclusion

The book set out to question the ways in which divisions in Ireland have been interpreted, and particularly the way the role of religion has been viewed. The focus of the debate was widened from Northern Ireland to take in the two dominant power alliances within the island as a whole and particular attention was paid to the perceptions, beliefs, and motivations which developed within the groups from the seventeenth century right up to and including the present day. Within this framework, religion has been shown to be of direct relevance to the underlying structural conflict. Contrary to the expectation that religion is increasingly irrelevant, it has appeared that, as a politically motivating power, religion exists in strength, entering into catholic–nationalist and protestant–loyalist common sense and dual hegemonies, and even the Southern state form.

Catholic nationalists and protestant loyalists are the real subjects of the battle for domination. The dominant cultures which solidify the internal class alliances are permeated by two types of opposing religious and nationalist/covenant ethos. The themes which traditionally differentiate Roman catholic and protestant world views come to the fore and take on crucial oppositional significance. Such conflict in Ireland was more evidently religious in past centuries, but not necessarily more religious. It is just that oppression used to be directly linked to religious truth in the abstract, as questions over the leadership of the church were coterminous with questions of political sovereignty. But incompatible forms of political religious power are still exercised by the two opposing alliances, though one only finds oppositional religious doctrine and ethical style stressed at the extremes. In the case of protestant fundamentalism, the stress is on salvation by faith, the evils of the catholic priesthood, and the mass. In the case of catholic monopoly, the stress is on a society governed according to God's naturally instituted laws, about which the catholic church is seen to have special insight and for which it has a guardian's role; to permit the infraction of such laws is seen as the beginning of the end of society.

The contents of the beliefs which hold sway within the protestant–loyalist alliance are clear, with now one now another theme dominating. Ulster protestants are a people with rights to an actual territory. For a significant and powerful number, protestant faith dominates political perceptions and, whatever the religious practice of the rank-and-file of the Orange order, the leadership of the order is either clerical or strongly church-going. Its self-justification is religious as well as religiously anti-catholic. For these Ulster protestants, loyalty to the monarchy has a strong religious element. The British monarch is seen as defender of protestant faith and rights against catholicism. The ideology of the Reformation has thus retained a significance right up to the present day. For this component of the protestant–loyalist alliance, the British government can and has been traitorous, betraying its covenant with the Northern Ireland people.

In conclusion we should ask if the strength of protestant–loyalist belief and its institutions are sufficient to maintain the alliance's solidarity. Protestant–loyalist ethnic identity is very limited in nature. It relies heavily on the opposing catholic–nationalist alliance for its maintenance. Whereas catholic nationalists would retain an albeit differently structured unity without the opposition of protestant loyalism, the reverse is not the case. The fundamental mode of opposition to catholic monopoly politics and Irish nationalism is essential to the solidarity of the protestant group, particularly in avoiding the development of a split or a choice between a covenant and a British identity. One must also remember that covenant perceptions are just as justifiable as nationalist ones, and that protestant loyalists could just as easily covenant with a European parliament as with a British one.

Social control within the protestant–loyalist alliance is aided by economic dependency on the United Kingdom. But the preservation of the group's domination in Ulster is likely to become the only source of solidarity if no agreement acceptable both to catholic nationalists and the British government is reached. The link with Britain would then become inessential. The continuing strength of protestant paramilitary groups and of protestant fundamentalism indicates a likely direction, if the power of the British-oriented groups were to prove ineffective. The setting up of a semi-autonomous protestant–loyalist statelet, probably functioning within some European community framework, might well emerge.

Though armed resistance would be difficult, they would probably be able to sustain control of a diminished territory. With the breaking of the union and no alternative offered to them by catholic nationalists other than a united Ireland, some of the values of liberty would also be lost to protestant loyalists as power shifted in the alliance. In the writer's view, the subsequent remix would become more rigid and fundamentalist, significantly less 'nice' than it has been in the recent past.

Protestant solidarity exists in Northern Ireland because protestant polity is under threat, and a part of that threat is monopoly catholicism. One of the aims of this book has been to point out that such a threat is real, and that Northern protestants would lose a number of their liberties in a united Ireland, unless monopoly catholicism in Ireland was ended and the church leadership modified its political–religious beliefs and strategy. Monopoly may not necessarily be an essential feature of catholic beliefs. There are other forms of political catholicism such as denominational catholicism in England and the USA, ecumenical catholicism in Holland, and liberation catholicism in Nicaragua. But monopoly catholicism has been the dominant form since the Middle Ages, and is a deep-seated characteristic of those Roman catholic national hierarchies which exist in countries populated by overwhelming catholic majorities. It dominates the Vatican administration even more strongly at the moment with the conservative pope, John-Paul II, and his close associate Cardinal Ratzinger. It also dominates large sectors of the Roman catholic population in France and Belgium, and conflicts with ecumenical catholicism in Holland and liberation catholicism in Nicaragua. What we are confronted with in Ireland is not that catholic nationalists are the cause of their own rigid forms of intellectual and popular religion, but that their own particular political problems of national emergence were bound up with the legacy of post-Tridentine religion. The harsher side of that religious form hits Ireland the hardest because the island had to become the political playground for the conflict between post-Reformation protestantism and cultural catholicism. Thus religion in part forms the conflict as much as it is formed by it. This indicates the ultimate paradox of the Irish catholic church which in so many ways is both progressive and incisive. While Irish men and women abroad in the third world labour for human rights and justice, such as in Central and South America, and the Philippines, and while even in Ireland justice and peace commissions seeks to counter the obvious violations of human rights in prisons and in the suspensions of normal juridical procedures, the routine, everyday structures of popular and clerical monopoly catholicism carry on regardless, contributing through a series of mediations to the conflict of the two opposing alliances.

References

Acquaviva, S. (1979), *The Decline of the Sacred in Industrial Society* (Oxford: Clarendon Press).

Adams, M. (1968), *Censorship: The Irish Experience* (Dublin: Scepter Books).

Adorno, T. W. (1950), *The Authoritarian Personality* (New York: Harper).

Akenson, D. H. (1970), *The Irish Education Experiment: The National System of Education in the Nineteenth Century* (London: Routledge & Kegan Paul).

——(1973), *Education and Enmity: The Control of Schooling in Northern Ireland 1920–1950* (Newton Abbott: David and Charles).

——(1975), *A Mirror to Kathleen's Face: Education in Independent Ireland, 1922–60* (Montreal: McGill and Queen's Univ. Press).

Althusser, L. (1971), *Lenin and Philosophy, and other Essays* (London: New Left Books).

Amin, S. (1982), *Class and Nation Historically and in the Current Crisis* (London: Heinemann).

Anderson, P. (1977), 'The Antinomies of Antonio Gramsci', *New Left Review*, 100: 5–78.

Arblaster, A. (1977), 'Britain in Ireland, Ireland in Britain', *Socialist Register 1977* (London: Merlin Press), 65–70.

Armstrong, D. and Saunders, H. (1985), *A Road too wide: The Price of Reconciliation in Northern Ireland* (Basingstoke: Marshall Morgan & Scott).

Armstrong, D. L. (1951), 'Social and Economic Conditions in the Belfast Linen Industry', *Irish Historical Studies*, 7: 235–69.

Arnold, B. (1984), *What Kind of Country? Modern Irish Politics 1968–1983* (London: Cape).

Arthur, P. (1984), *Government and Politics of Northern Ireland* (London: Longman).

Aspects of Catholic Education (1970), Papers read at a conference organized by the Guild of Catholic Teachers, Diocese of Down and Conor, 21–2 Mar. 1970, St Joseph's College of Education, Belfast.

Aunger, E. A. (1983), 'Religion and Class: An Analysis of 1971 Census Data', in R. J. Cormack and R. D. Osborne (eds.), *Religion, Education and Unemployment* (Belfast: Appletree).

Ayearst, M. (1971), *The Republic of Ireland: Its Government and Politics* (London: Univ. of London Press).

BAILIE, W. D. (1981), 'The Reverend Samuel Barber 1738–1811: National Volunteer and United Irishman', in Haire, *et al.* 1981: 72–95.

BAKER, D. (ed.) 1975, *Church Society and Politics*, Studies in Church History, 12 (Oxford: Blackwell).

BALL, S. (1981), *Beachside Comprehensive* (Cambridge: CUP).

Ballymascanlon Working Party IV (1975), 'Christianity and Secularism': Final Report of Working Party IV presented to the 3rd Inter-Church Meeting to be held at Ballymascanlon, Dundalk on 23 Apr. 1975, Church House, Belfast.

BANTON, M. (1987), *Racial Theories* (Cambridge: CUP).

BARKLEY, J. M. (1959), *A Short History of the Presbyterian Church in Ireland* (Belfast: Presbyterian Church in Ireland).

—— (1963), *The Eldership in Irish Presbyterianism* (Belfast: publisher not given).

—— (1966), *Presbyterianism* (Belfast: General Assembly of the Presbyterian Church in Ireland).

—— (1968), *The Romeward Trend in Irish Presbyterianism* (Belfast: The Presbyterian College).

—— (1970), 'Anglican Presbyterian Relations', in Hurley 1970.

—— (1972), 'The Arian Schism in Ireland, 1830', in D. Baker (ed.), *Schism, Heresy and Religious Protest* (Cambridge: CUP), 323–39.

—— (1981), 'The Presbyterian Minister in Eighteenth Century Ireland', in Haire, *et al.* 1981: 46–71.

BARRINGTON, D. (1959), *The Church, the State and the Constitution* (Dublin: Catholic Truth Society of Ireland).

BARRINGTON, R. (1987), *Health, Medicine and Politics in Ireland 1900–1970* (Dublin: Institute of Public Administration).

BARRITT, D. P. (1982), *Northern Ireland: A Problem to every Solution* (London: Quaker Peace and Service in association with Friends Peace, Leeds).

—— and BOOTH, A. (1972), *Orange and Green*, 3rd edn. (Sedbergh: Society of Friends).

—— and CARTER, C. F. (1972), *The Northern Ireland Problem: A Study in Group Relations*, 2nd edn. (London: OUP).

BARRY, J. G. (ed.) (1974), *Historical Studies*, Papers read before the Irish Conference of Historians IX, Cork, 29–31 May 1971 (Belfast: Blackstaff).

BARTH, F. (ed.) (1970), *Ethnic Groups and Boundaries* (London: Allen & Unwin).

BARTLETT, T., CURTIN, C., O'DWYER, R., and O'TUATHAIGH, G. (eds.) (1988), *Irish Studies: An Introduction* (Dublin: Gill and Macmillan).

BEACH, S. W. (1977), 'Religion and Political Change in Northern Ireland', *Sociological Analysis*, 38: 37–48.

BEAMES, M. R. (1982), 'The Ribbon Societies: Lower-Class Nationalism in Pre-Famine Ireland', *Past and Present*, 97: 128–43.
BECKETT, J. C. (1969), *The Making of Modern Ireland 1603–1923* (London: Faber & Faber).
—— (1976), *The Anglo-Irish Tradition* (Belfast: Blackstaff Press).
BELL, G. (1976), *The Protestants of Ulster* (London: Pluto Press).
BELLAH, R. N. (1970), *Beyond Belief* (New York: Harper & Row).
BENDER, L. (1960), *Ius Publicum Ecclesiasticum* (Hilversum: Paulus Brand).
BENSON, P. L. and WILLIAMS, D. L. (1982), 'Religion on Capitol Hill: Myths and Realities' (San Francisco, Calif.: Harper & Row).
BERGER, P. L., BERGER, B., and KELLNER, H. (1974), *The Homeless Mind* (Harmondsworth: Penguin).
BERNSTEIN, B. (1975), *Class Codes and Control*, iii (London: Routledge & Kegan Paul).
BEW, P. (1979), *Land and the National Question in Ireland 1858–82* (Atlantic Highlands, NJ: Humanities Press).
—— GIBBON, P., and PATTERSON, H. (1979), *The State in Northern Ireland* (Manchester: Manchester Univ. Press).
—— —— —— (1980), 'Aspects of Nationalism and Socialism in Ireland: 1969–78', in Morgan and Purdie 1980: 152–71.
—— and PATTERSON, H. (1985), *The British State and the Ulster Crisis from Wilson to Thatcher* (London: Verso).
BHALDRAITHE, E. DE (1971), 'Joint Pastoral Care of Mixed Marriages', *Furrow*, 22: 124–33.
—— (1981), 'The Ecumenical Marriage', *Furrow*, 32: 639–48.
BINCHY, W. (1984), *Is Divorce the Answer?* (Blackrock, Co. Dublin: Irish Academic Press).
BIRRELL, W. D. and ROCHE, D. J. D. (1977), 'The Political Role and Influence of Clergymen in Northern Ireland', in Sociological Association of Ireland, Proceedings of 1st and 4th Annual Conferences, Queen's Univ., Belfast, 44–8.
BOLAND, K. (1982), *The Rise and Decline of Fianna Fáil* (Cork and Dublin: Mercier).
—— (1984), *Fine Gael: British or Irish?* (Cork and Dublin: Mercier).
BOSERUP, A. (1972), 'Contradictions and Struggles in Northern Ireland', *Socialist Register 1972* (London: Merlin Press), 157–93.
BOWEN, D. (1970), *Souperism: Myth or Reality?* (Cork: Mercier).
—— (1978), *The Protestant Crusade in Ireland* (Dublin: Gill and Macmillan).
—— (1983), *Paul Cardinal Cullen and the Shaping of Modern Irish Catholicism* (Dublin: Gill and Macmillan).
BOWEN, K. (1983), *Protestants in a Catholic State: Ireland's Privileged Minority* (Dublin: Gill and Macmillan).

BOWYER-BELL, J. (1980), *The Secret Army: The I.R.A.* (London: Fontana).
BOYCE, D. G. (1982), *Nationalism in Ireland* (Dublin: Gill and Macmillan).
BOYD, A. (1969), *Holy War in Belfast: A History of the Troubles in Northern Ireland* (Tralee: Anvil Books).
BOYD, R. (1986), 'Peace and Reconciliation Institutions in Ireland', *Ireland Today*, 1026: 11–13.
BOYLE, J. W. (1962), 'Belfast Protestant Association and the Independent Orange Order, 1901–10', *Irish Historical Studies*, 13: 117–52.
BOYLE, K., HADDEN, T., and HILLYARD, P. (1975), *Law and the State: The Case of Northern Ireland* (London: Martin Robertson).
—— —— —— (1980), *Ten Years on in Northern Ireland: The Legal Control of Political Violence* (London: Cobden Trust).
—— —— (1985a), *Ireland: A Positive Proposal* (Harmondsworth: Penguin).
—— —— (1985b), 'Is the Agreement Viable without the Unionists?' *Fortnight*, 230: 3–4.
—— —— (1985c), 'A Positive Proposal for a New Anglo-Irish Treaty on Northern Ireland', *Fortnight*, 224: 8–9.
BRENNAN, N. (1973), 'Catechetics or Religious Education in Post-Primary Schools', unpub. Paper, Mt Oliver, Dundalk.
BRESLIN, A. and WEAFER, J. (1985), *Religious Beliefs, Practice and Moral Attitudes: A Comparison of Two Irish Surveys 1974–1984* (Maynooth, Ire.: Council for Research and Development, Report No. 21).
BREUILLY, J. (1985), *Nationalism and the State* (Manchester: Manchester Univ. Press).
British and Irish Communist Organization (1971), *On the Democratic Validity of the Northern Ireland State* (Belfast: BICO).
—— (1972), *The Economics of Partition* (Belfast: BICO).
—— (1973), *The Birth of Ulster Unionism* (Belfast: BICO).
BRONFENBRENNER, U. (1972), 'Another World of Childhood', *New Society*, 19/489: 279–80.
BROOKE, P. (1987), *Ulster Presbyterianism in Historical Perspective 1610–1970* (Dublin: Gill and Macmillan).
—— and ROBINSON, C. (1984), *What Happened on the Twelfth?* (Belfast: Athol Press).
BROWN, T. (1981), *Ireland: A Social and Cultural History 1922–1979* (London: Fontana).
BRUCE, S. (1984), *Firm in the Faith: The Survival of Conservative Protestantism* (Aldershot: Gower).
—— (1986a), 'A House Divided: Protestant Schisms and the Rise of Religious Tolerance', *Sociological Analysis*, 47/1: 21–8.
—— (1986b), *God Save Ulster: The Religion and Politics of Paisleyism* (Oxford: OUP).

BUCI-GLUCKSMANN, C. (1980), *Gramsci and the State* (London: Lawrence and Wishart).
BUCKLAND, P. (1973a), *Irish Unionism* (London: Historical Assoc.).
——(1973b), *Irish Unionism, ii, Ulster Unionism and the Origins of Northern Ireland 1886–1922* (Dublin: Gill and Macmillan).
——(1979), *The Factory of Grievances: Devolved Government in Northern Ireland 1921–39* (Dublin: Gill and Macmillan).
BUCKLEY, A. D. (1984), 'Walls within Walls: Religion and Rough Behaviour in an Ulster Community', *Sociology*, 18/1: 19–32.
BURKE, C. (1984), 'Appointing a Teacher: Religious Requirements', *Furrow*, 35: 246–51.
BURTON, F. (1978), *The Politics of Legitimacy: Struggles in a Belfast Community* (London: Routledge & Kegan Paul).
CALVIN, J. (1960), *Institutes of the Christian Religion*, 2 vols., ed. J. T. McNeil, trans. F. L. Battles (Philadelphia, Pa., Westminster Press).
CAMPBELL, J. J. (1964), *Catholic Schools: A Survey of a Northern Ireland Problem* (Dublin: Fallons).
CANNY, N. (1982), 'The Formation of the Irish Mind: Religion, Politics and Gaelic-Irish Literature 1580–1750', *Past and Present*, 95: 91–116.
CARSON, J. T. (1958), *God's River in Spate: The Story of the Religious Awakening in Ulster in 1859* (Belfast: publisher not given).
Census of Northern Ireland (1961, 1971, 1981) (Belfast: Northern Ireland Office).
Census of Population (1971), *Census of Population of Ireland 1971*, i–xiv (Dublin: Central Statistics Office).
——(1979), *Census of Population of Ireland 1979*, i, ii, iii (Dublin: Central Statistics Office).
——(1981), *Census of Population of Ireland 1981, Preliminary Report* (Dublin: Central Statistics Office).
CHUBB, B. (1982), *The Government and Politics of Ireland*, 2nd edn. (London: Longman).
Church of Ireland General Convention (1870), *Journal of the General Convention* (Dublin: Church Representative Body).
Church of Ireland General Synod (1871–1984), *Journal of the General Synod* (Dublin: Representative Church Body).
CLANCY, P., DRUDY, S., LYNCH, K., and O'DOWD, L. (eds.) (1986) *Ireland: A Sociological Profile* (Dublin: Institute of Public Administration).
CLARKE, S. (1979), *Social Origins of the Irish Land War* (Princeton, NJ: Princeton Univ. Press).
CLARKE, D. M. (1985), *Church and State* (Cork: Univ. Press).
Code of Canon Law (1917), *Codex Iuris Canonici* (Rome: Vatican). Translations taken from S. Woywood, *A Practical Commentary on the Code of Canon Law* (New York: Wagner Inc., 1957).

Code of Canon Law (1983), *The Code of Canon Law* (London: Collins).
COHEN, A. P. (ed.) (1982), *Belonging: Identity and Social Organization in British Rural Cultures* (Manchester: Manchester Univ. Press).
COHEN, R. (1978), 'Ethnicity: A Problem and Focus in Anthropology', *Annual Review of Anthropology*, 7: 379–403.
COHEN, S. (1972), *Folk Devils and Moral Panics: The Creation of the Mods and Rockers* (London: MacGibbon and Kee).
COMAN, P. (1983), *Roman Catholic Social Teaching, Conservatism, and Sociological Theory* (Leeds: Trinity and All Saints College).
COMPTON, P. (1978), *Northern Ireland: A Census Atlas* (Dublin: Gill and Macmillan).
—— (1981), 'The Demographic Background', Watt 1981: 74–92.
—— (1985), 'An Evaluation of the Changing Religious Composition of the Population of Northern Ireland', *Economic and Social Review*, 16: 201–24.
—— and POWER, J. (1986), 'Estimates of the Religious Composition of Northern Ireland Local Government Districts in 1981 and Change in the Geographical Pattern of Religious Composition between 1971 and 1981', *Economic and Social Review*, 17/1: 87–105.
—— and COWARD, J. (1989), *Fertility and Family Planning in Northern Ireland* (Aldershot: Avenbury Press).
CONNELL, K. H. (1950), *The Population of Ireland* (Oxford: Clarendon Press).
—— (1968), *Irish Peasant Society: Four Historical Essays* (Oxford: OUP).
CONNICK, A. J. (1964), 'Canonical Doctrine Concerning Mixed Marriages—before Trent and during the Seventeenth and Eighteenth Centuries', *Jurist*, 20: 295–326, 398–418.
CONNOLLY, J. (1948), *Socialism and Nationalism* (Dublin: Blackstaff Press).
CONNOLLY, S. (1985), *Religion and Society in Nineteenth Century Ireland* (Dundalk: Dundalgan Press).
CONNOLLY, S. J. (1981), 'Catholicism in Ulster, 1800–1850', in P. Roebuck (ed.), *Plantation to Partition: Essays in Ulster History in Honour of J. L. McCracken* (Belfast: Blackstaff Press), 157–71.
—— (1982), *Priests and People in Pre-Famine Ireland* (Dublin: Gill and Macmillan).
CONNOLLY, W. (ed.) (1984), *Legitimacy and the State* (Oxford: Blackwell).
Constitution of the Irish Free State (1922), 'Draft Constitution of the Irish Free State', Parliamentary Papers, 1922, xvii. 607–23 (London: HMSO).
Constitution of Ireland (1937) (Dublin: Government Publications).
CONWAY, Card. W. (1970), *Catholic Schools* (Dublin: Catholic Communications Institute of Ireland).

—— (1972), Interview in *Newsweek* (European edn.), 4 Sept., p. 56.
COOGAN, T. P. (1980), *The I.R.A.* (London: Fontana).
COONEY, J. (1986), *The Crozier and the Dáil, Church and State 1922–1986* (Cork and Dublin: Mercier).
CORISH, P. (1985), *The Irish Catholic Experience: A Historical Survey* (Dublin: Gill and Macmillan).
CORISH, P. J. (1983), 'Paul Cardinal Cullen and the Shaping of Modern Irish Catholicism', *Furrow*, 34/12: 791–3.
CORKEY, Revd W. (1911), *The McCann Mixed Marriage Case* (Edinburgh: The Knox Club).
CORMACK, R. J. and OSBORN, R. D. (eds.) (1983), *Religion, Education and Employment: Aspects of Equal Opportunity in Northern Ireland* (Belfast: Appletree Press).
COSTELLO, D. (1956), 'The Natural Law and the Irish Constitution', *Studies*, 45: 404–14.
COVENTRY, J. (1975), 'Positions and Trends in Britain', Hurley 1970: 115–19.
COX, W. H. (1985), 'Who Wants a United Ireland?' *Government and Opposition*, 20: 29–47.
CRAWFORD, R. G. (1981), 'The Second Subscription Controversy and the Personalities of the Non-Subscribers', Haire, *et al.* 1981: 96–115.
CRAWFORD, V. (1933), *Catholic Social Doctrine 1891–1931* (Oxford: Catholic Social Guild).
CROTTY, R. D. (1966), *Irish Agricultural Production: Its Volume and Structure* (Mystic, Conn.: Lawrence Verry Inc.).
CULLEN, L. M. (1972), *An Economic History of Ireland since 1660* (London: Batsford).
—— (1983), *The Emergence of Modern Ireland 1600–1900* (Dublin: Gill and Macmillan).
CUNNINGHAM, T. P. (1964), 'Mixed Marriages in Ireland before *Ne Temere*', *Irish Ecclesiastical Record*, 101: 53–6.
CURTIN, C., KELLY, M., and O'DOWD, L. (eds.) (1983), *Culture and Ideology in Ireland* (London: Marion Boyars Ltd.).
CURTIS, Jr., L. P. (1968), *Anglo-Saxons and Celts: A Study in Anti-Irish Prejudice in Victorian England* (Bridgeport: Univ. of Connecticut).
DALY, Bp., C. (1972), 'On Obligation of State to Prevent Contraception and Divorce', *Irish Times*, 12 Aug.
—— (1977), 'The Future of Christianity in Ireland', in *Doctrine and Life* (spec. issue), Mar. and Apr. 1977, pp. 105–24.
—— (1978), 'Separate Schools: Distinctiveness without Divisiveness', Paper read at Magee College seminar, Derry.
DARBY, J. (1976), *Conflict in Northern Ireland* (Dublin: Gill and Macmillan).

DARBY, J. (ed.) (1983), *Northern Ireland: The Background to the Conflict* (Belfast: Appletree Press).

—— (1987), *Intimidation and Control of Conflict in Northern Ireland* (Syracuse, NY: Syracuse Univ. Press).

—— MURRAY, D., BATTS, D., DUNN, S., FARREN, S., and HARRIS, J. (1977), *Education and Community in Northern Ireland: Schools Apart?* (Coleraine: New Univ. of Ulster).

DARBY, J. P. (1974), 'Segregated Schooling—Does it Contribute to Sectarianism?' in *Sectarianism*, Papers of the 22nd Annual Summer School of the Social Study Conference, Dublin, 59–68.

DEANE, D. (1974), 'Mixed Marriage: Irish and Swiss Bishops' Statements Compared', *Furrow*, 25/10: 544–8.

DENZINGER, H. and SCHÖNMETZER, A. (eds.) (1973), *Enchiridion Symbolorum Definitionum et Declarationum de Rebus Fidei et Morum* (New York: Herder).

DE TOCQUEVILLE, A. (1957), *Journeys to England and Ireland*, ed. J. P. Mayer (London: Faber & Faber).

DONALD, J. and HALL, S. (eds.) (1985), *Politics and Ideology* (Milton Keynes: Open Univ. Press).

DONNAN, H. and MCFARLANE, G. (1986), 'Social Anthropology and the Sectarian Divide in Northern Ireland', Jenkins, Donnan, and McFarlane 1986: 23–37.

DONNELLY, J. (1975), *The Land and the People of Nineteenth Century Cork: The Rural Economy and the Land Question* (London: Routledge & Kegan Paul).

DOUGLAS, M. (1971), *Purity and Danger* (Harmondsworth: Penguin).

DOWLING, P. J. (1968), *The Hedge Schools of Ireland* (Cork: Mercier).

DOYLE, T. P. (1985), 'The R.C. Church and Mixed Marriages', *Ecumenical Trends* (Greymoor), 14/6.

DRUDY, J. (ed.) (1982), *Ireland: Land, Politics and People*, Irish Studies, ii (Cambridge: CUP).

DUBY, G. (1978), *Medieval Marriage: Two Models from Twelfth-Century France* (London: Johns Hopkins Univ. Press).

—— (1984), *The Knight, the Lady and the Priest* (London: Allen Lane).

DUNSTAN, G. (1975), 'Appendix: Natural and Sacramental Marriage', Hurley 1975: 67–72.

EASTHOPE, G. (1976), 'Religious War in Northern Ireland', *Sociology*, 10: 427–50.

Ecumenical Notes (1976), 'Ecumenical Notes', *One in Christ* (1976): 269–80.

EDWARDS, O. D. (1970), *The Sins of our Fathers* (Dublin: Gill and Macmillan).

EGGLESTON, J. (ed.) (1974), *Contemporary Research in the Sociology of Education* (London: Methuen).

EIPPER, C. (1985), *The Ruling Trinity: A Community Study of Church, State and Business in Ireland* (London: Pluto).

ELLIOTT, E. P. M. (1978), 'The Church of Ireland and Northern Ireland Schools', *Search* (Ireland), 1: 31–9.

EVERSLEY, D. (1989), *Religion and Employment in Northern Ireland* (London: Sage).

FANON, F. (1961), *Les Damnés de la terre*, Preface by J.-P. Sartre (Paris: Maspero).

FARRELL, B. (1970–1971), 'The Drafting of the Irish Free State Constitution', *Irish Jurist*, 5: 115–40; 6: 111–35, 343–56.

—— (1973), 'The Paradox of Irish Politics' in Farrell (ed.), *The Irish Parliamentary Tradition* (Dublin: Gill and Macmillan).

FARRELL, M. (1976), *Northern Ireland: The Orange State* (London: Pluto Press).

—— (1977), 'Northern Ireland—An Anti-Imperialist Struggle', *Socialist Register 1977* (London: Merlin Press), 71–80.

—— (1983), *Arming the Protestants* (London: Pluto Press).

FAUGHNAN, S. (1988), 'The Jesuits and the Drafting of the Irish Constitution of 1937', *Irish Historical Studies*, 26: 79–102.

FAUL, D. (1977), 'Why Catholic Schools in Ireland', *Furrow*, 28/1: 18–21.

FITZGERALD, G. (1964), 'Towards a National Purpose', *Studies* (Winter Issue).

—— (1972), 'Towards a New Ireland' (Dublin: Charles Knight and Co.).

—— (1975), 'Address to the International Consultation on Mixed Marriage', Hurley 1975: 188–93.

FOGARTY, M., RYAN, L., and LEE, J. (1984), *Irish Values and Attitudes: The Irish Report of the European Values Systems Study* (Dublin: Dominican Press).

FORD, R. (1986), 'Mapping out a New Ulster', *The Times*, 9 May, 16.

FRANK, A. G. (1971), *Capitalism and Under-development in Latin-America* (New York: Monthly Review Press).

FRASER, M. (1974), *Children in Conflict* (Harmondsworth: Penguin).

FREEMANTLE, A. (ed.) (1956), *The Papal Encyclicals* (New York: New American Library of World Literature Inc.).

FULTON, J. (1975), 'Intermarriage and the Irish Clergy: A Sociological Study', in Hurley 1975.

—— (1977), 'Is the Irish Conflict Religious? A Sociology for the Irish Churches', *Social Studies*, 5: 191–201.

—— (1981*a*), 'Experience, Alienation and the Anthropological Condition of Religion', *Annual Review of the Social Sciences of Religion*, 5: 1–32.

—— (1981*b*), 'The Debate in Denominational and Integrated Schooling in Ireland', Institute for the Study of Worship, Birmingham Univ., *Research Bulletin*, 102–14.

FULTON, J. (1986), 'State, Religion and Law in Ireland', in T. Robbins and R. Robertson (eds.), *Church–State Relations: Tensions and Transitions* (New Brunswick and London: Transaction Books).
—— (1987*a*), 'Religion as Politics in Gramsci', *Sociological Analysis*, 48: 197–216.
—— (1987*b*), 'The Historical and Social Ground of Religion and Conflict in Modern Ireland: A Critical, Holistic Approach', Ph.D. thesis, LSE.
—— (1988), 'Sociology, Religion and "the Troubles" in Northern Ireland: A Critical Approach', *Social and Economic Review*, 20/1.
GAINE, M. (1968), 'The Development of Official Roman Catholic Educational Policy in England and Wales', Jebb 1968: 137–64.
GALLAGHER, E. and WORRALL, S. (1982), *Christians in Ulster 1968–1980* (London: OUP).
GALLAGHER, M. (1985), *Political Parties in the Republic of Ireland* (Dublin: Gill and Macmillan).
—— (1989), 'Irish Political Data, 1988', *Irish Political Studies*, 4: 145–61.
GALLAGHER, T. and O'CONNELL, J. (eds.) (1983), *Contemporary Irish Studies* (Manchester: Manchester Univ. Press).
GARVIN, T. (1981), *The Evolution of Irish National Politics* (Dublin: Gill and Macmillan).
—— (1986), 'Priests and Patriots: Irish Separatism and Fear of the Modern 1890–1914', *Irish Historical Studies*, 25: 67–81.
GEERTZ, C. (1966), 'Religion as a Cultural System', Banton 1987: 1–46.
—— (1971), *Islam Observed* (London: Univ. of Chicago Press).
GELLNER, E. (1964), *Thought and Change* (Chicago, Ill.: Univ. of Chicago Press).
—— (1983), *Nations and Nationalism* (London: Blackwell).
GERTH, H. H. and MILLS, C. W. (eds.) (1946), *From Max Weber* (Oxford: OUP).
GIBBON, P. (1975), *The Origins of Ulster Unionism* (Manchester: Manchester Univ. Press).
GRAMSCI, A. (1971), *Selections from the Prison Notebooks*, ed. Q. Hoare and G. N. Smith (London: Lawrence and Wishart).
—— (1975), *Quaderni del Carcere*, critical edn. of the Istituto Gramsci, ed. V. Gerratana (Turin: Einaudi).
—— (1977), *Il Vaticano e l'Italia*, 3rd edn. (Rome: Editori Riuniti).
GREELEY, A. M. (1976), *The American Catholic: A Social Portrait* (New York: Basic Books).
—— and ROSSE, P. H. (1966), *The Education of Catholic Americans* (New York: Doubleday).
—— MCCREADY, W. C., and MCCOURT, K. (1976), *Catholic Schools in a Declining Church* (Kansas City: Sheed and Ward).

—— Saldanah, S., McCready, W., and McCourt, K. (1975), 'American Catholics: Ten Years Later', *Critic* (Jan. 1975), 37–47.

Greer, J. (1972), 'Mixed Marriages in Ireland', *Doctrine and Life*, 22/9: 491–4.

Greer, J. E. (1975), 'Religious Education: Contemporary Approaches', Paper read to ACT seminar, Presbyterian Centre, Belfast.

—— (1980–1), 'The Persistence of Religion in Northern Ireland', *Social Studies*, 6/4: 317–34.

Grogan, J. (1954), 'The Constitution and the Natural Law', *Christus Rex*, 8: 201–18.

Guild of Catholic Teachers (1971), 'Aspects of Catholic Education', Papers read at a Conference organized by the Guild of Catholic Teachers Diocese of Down and Conor, 21–31 Mar. 1970, St Joseph's College of Education, Belfast, 1971.

Haire, J. L. M., et al. (1981), *Challenge and Conflict: Essays in Irish Presbyterian History and Doctrine* Antrim: W. G. Baird Ltd.).

Hall, S., et al. (1978), *Policing the Crisis: Mugging, the State and Law and Order* (London: Macmillan).

—— (1979), 'Politics and Ideology: Gramsci', in *On Ideology*, Birmingham (GB), Centre for Cultural Studies (London: Hutchinson).

Hannon, D. (1979), *Displacement and Development: Class, Kinship and Social Change in Irish Rural Communities*, ESRI Paper No. 96.

Harbison, J. and Harbison, J. (eds.) (1980), *A Society under Stress: Children and Young People in Northern Ireland* (Shepton Mallet: Open Books).

Harris, R. (1972), *Prejudice and Tolerance in Ulster: A Study of Neighbours and 'Strangers' in a Border Community* (Manchester: Manchester Univ. Press).

Hechter, M. (1975), *Internal Colonialism* (London: Routledge & Kegan Paul).

Hempton, D. N. (1980), 'The Methodist Crusade in Ireland 1795–1845', *Irish Historical Studies*, 22/85: 33–48.

Hempton, D. (1984), *Methodism and Politics in British Society 1750–1850* (London: Hutchinson).

Hepburn, A. C. (1980), *The Conflict of Nationality in Modern Ireland* (London: Edward Arnold).

Heron, A. (1975), 'The Ecclesiological Problems of Interchurch Marriage', Hurley 1975: 73–103.

Heskin, K. (1980), *Northern Ireland: A Psychological Analysis* (Dublin: Gill and Macmillan).

Heslinga, M. W. (1962), *The Irish Border as a Cultural Divide* (Assen: van Gorcum).

Hickey, J. (1981), 'Religion, Values and Daily Life: A Case Study in

Northern Ireland', CISR, 16th Annual Conference, Lausanne, 301–16.
—— (1984), *Religion and the Northern Ireland Problem* (Dublin: Gill and Macmillan).
HILLYARD, P. (1983), 'Law and Order', Darby 1983: 32–60.
—— and BOYLE, K. (1982), 'The Diplock Court Strategy: Some Reflections on Law and the Politics of Law', Kelly, O'Dowd, and Wickham, 1982: 1–20.
HOLMES, R. F. G. (1981), 'Controversy and Schism in the Synod of Ulster in the 1820s', Haire, *et al.* 1981: 116–33.
HOPPEN, K. T. (1984), *Elections, Politics and Society in Ireland 1832–1885* (Oxford: OUP).
HOPPL, H. (1982), *The Christian Polity of John Calvin* (Cambridge: CUP).
HORNSBY-SMITH, M. P. (1987), *Roman Catholics in England and Wales* (Cambridge: CUP).
HUNTER, J. (1982), 'An Analysis of the Conflict in Northern Ireland', Rea, 1982: 9–59.
HURLEY, M. (ed.) (1970), *Irish Anglicanism 1869–1969* (Dublin: Allen Figgis).
—— (ed.) (1975), *Beyond Tolerance: The Challenge of Mixed Marriage* (Dublin: Chapman).
HYNES, E. (1978), 'The Great Hunger and Irish Catholicism', *Societas*, 8: 137–56.
INGLIS, T. (1987), *Moral Monopoly: The Catholic Church in Modern Irish Society* (Dublin: Gill and Macmillan).
Inter-Church Group on Faith and Politics (1986), *Breaking Down the Enmity: Faith and Politics in Northern Ireland*, 2nd edn. (Dublin: Inter-Church Group on Faith and Politics).
International Union of Social Studies (1929), *Code of Social Principles*, 1st edn. (Malines).
—— (1937), *Code of Social Principles*, 2nd edn. (Oxford: Catholic Social Guild).
Ireland Today (1949–), Journal for the Department of Foreign Affairs, Republic of Ireland (Dublin: Dept. of Foreign Affairs).
Irish Ecumenical News (1982–), Periodical, Irish School of Ecumenics, Dublin.
Irish Episcopal Conference (1970), 'Mixed Marriages', *Furrow*, 21: 732–3.
—— (1976), *Directory on Ecumenism in Ireland* (Dublin: Veritas).
—— (1980), 'Conscience and Morality', Doctrinal Satement (Dublin: Irish Messenger Publication).
—— (1983*a*), *The Storm that Threatens: War and Peace in the Nuclear Age* (Dublin: Veritas).
—— (1983*b*), *Directory on Mixed Marriages* (Dublin: Veritas).
—— (1983*c*), *Preparing for a Mixed Marriage* (Dublin: Veritas).

—— (1984), *Submission to the New Ireland Forum* (Dublin: Veritas).
—— (1985), *Love is for Life* (Dublin: Veritas).
—— (1986), *Marriage, the Family and Divorce* (Dublin: Veritas).
JEBB, P. (ed.) (1968), *Religious Education: Drift or Decision?* (London: Longman, Darton and Todd).
JENKINS, R. (1969), 'Religious Conflict in Northern Ireland', in D. Martin (ed.), *A Sociological Yearbook of Religion in Britain* (London: SCM Press).
—— (1983), *Lads, Citizens and Ordinary Kids: Working-Class Life-Styles in Belfast* (London: Routledge & Kegan Paul).
—— (1984*a*), 'Understanding Northern Ireland', *Sociology*, 18: 253–64.
—— (1986), 'Northern Ireland: In What Sense "Religions" in Conflict?' Jenkins, Donnan and McFarlane 1986: 1–22.
—— DONNAN, H., and McFARLANE, G. (1986), *The Sectarian Divide in Northern Ireland Today*, Royal Anthropological Institute of Great Britain and Northern Ireland, London, Occasional Paper No. 41.
JESSOP, B. (1980), 'On Recent Marxist Theories of Law, the State and Juridico-Political Ideology', *International Journal of the Sociology of Law*, 8: 339–68.
—— (1982), *The Capitalist State: Marxist Theories and Methods* (Oxford: M. Robertson).
JOHN PAUL II (1979), *The Pope in Ireland* (Dublin: Veritas).
JOLL, J. (1977), *Gramsci* (London: Fontana).
KAVANAGH, J. (1954), 'Emigration: A Letter', *Bell*, 19/10: 54–8.
KEDOURIE, E. (1985), *Nationalism*, rev. edn. (London: Hutchinson).
KEE, R. (1972), *The Green Flag: A History of Irish Nationalism* (London: Weidenfeld & Nicolson).
KEENAN, D. (1983), *The Catholic Church in Nineteenth Century Ireland: A Sociological Study* (Dublin: Gill and Macmillan).
KELLEY, J., and McALLISTER, I. (1984), 'The Genesis of Conflict: Religion and Status Attainment in Ulster, 1968', *Sociology*, 18: 171–90.
KELLEY, J. M. (1967), *Fundamental Rights in the Irish Law and Constitution* (Dublin: Gill and Macmillan).
KELLY, M., O'DOWD, L., and WICKHAM, J. (eds.) (1982), *Power, Conflict and Inequality* (Boston, Mass., and London: Marion Boyars Ltd.).
KEOGH, D. (1986), *The Vatican, The Bishops and Irish Politics 1919–39* (Cambridge: CUP).
—— (1987), 'The Constitutional Revolution: An Analysis of the Making of the Constitution', *Administration*, 35: 4–84.
KILLEN, W. D. (1853), *History of the Presbyterian Church in Ireland*, iii. *1700–1853* (London: Whittaker & Co.).
—— (1875), *Ecclesiastical History of Ireland*, 2 vols. (London: pub. not given).

LAFFAN, M. (1983), *The Partition of Ireland 1911–25* (Dundalk: Dungaldan).
LARKIN, E. (1975), *The Roman Catholic Church and the Creation of the Modern Irish State 1878–1886* (Philadelphia, Pa.: Univ. of Philadelphia Press).
——(1976), *The Historical Dimensions of Irish Catholicism*, Essays previously published in the *American Historical Review*, 1967–75 (New York: American Historical Review Publications).
——(1978), *The Roman Catholic Church and the Plan of Campaign in Ireland 1886–1888* (Cork: Cork Univ. Press).
——(1979), *The Roman Catholic Church in Ireland and the Fall of Parnell 1888–1891* (Liverpool: Liverpool Univ. Press).
——(1980), *The Making of the Roman Catholic Church in Ireland 1850–60* (Chapel Hill: Univ. of North Carolina Press).
LA ROCCA, T. (1981), *Gramsci e la Religione* (Brescia: Queriniana).
LATIMER, W. T. (1902), *History of the Irish Presbyterians* (Belfast: Cleeland).
LECKY, W. E. H. (1912), *History of Ireland in the Eighteenth Century*, 5 vols. (London: Longmans, Green and Co.).
LEE, G. (1969), 'Aspects of the Irish Constitution', *Christus Rex*, 23: 65–70.
LEE, J. (1973), *The Modernisation of Irish Society 1848–1918* (Dublin: Gill and Macmillan).
LEE, R. M. (1981), 'Interreligious Courtship and Marriage in Northern Ireland', Ph. D. thesis, Univ. of Edinburgh.
——(1985), 'Intermarriage, Conflict and Social Control in Ireland: The Decree "Ne Temere"', *Economic and Social Review*, 17/1: 11–27.
LENNON, B. D. (1980*a*), 'Two Church Marriages in Ireland: An Ecclesiological Problem', MA diss., Univ. of Hull/Irish School of Ecumenics, Dublin.
——(1980*b*), 'Interchurch Marriages: Torn between Divided Churches', *Furrow*, 31: 309–21.
LEO XIII (1903), *The Great Encyclical Letters of Pope Leo XIII* (New York: Benzinger).
LÉVI-STRAUSS, C. (1968), *Structural Anthropology*, i. (New York: Basic Books).
LEYTON, E. (1966), 'Conscious Models and Dispute Regulations in an Ulster Village', *Man* (NS), 1: 534–42.
——(1975), 'Opposition and Integration in Ulster', *Man* (NS), 9: 185–98.
LIJPHART, A. (1975), 'Review Article: The Northern Ireland Problem: Cases, Theories and Solutions', *British Journal of Political Science*, 5/1: 83–106.
LONGFORD, Earl of, and O'NEILL, T. P. (1970), *Eamon de Valera* (Dublin: Gill and Macmillan).

Lyons, F. S. L. (1973), *Ireland since the Famine*, rev. edn. (London: Collins/Fontana).

—— (1982), *Culture and Anarchy in Ireland* (Oxford: OUP).

McAllister, I. (1980), *Territorial Differentiation and Party Development in Northern Ireland* (Glasgow, Univ. of Strathclyde: Centre for the Study of Public Policy, Studies in Public Policy No. 66).

—— (1982), 'The Devil, Miracles and the Afterlife: The Political Sociology of Religion in Northern Ireland', *British Journal of Sociology*, 33/3: 330–47.

—— (1983a), 'Class, Region, Denomination and Protestant Politics in Ulster', *Political Studies*, 41: 275–83.

—— (1983b), 'Religious Commitment and Social Attitudes in Ireland', *Economic and Social Review*, 14: 185–202.

McCaffery, L. J. (1973), 'Irish Nationalism and Irish Catholicism: A Study of Cultural Identity', *Church History*, 42: 524–34.

—— (1976), *Irish Nationalism and the American Contribution* (Salem, NH: Ayer Publishing Co.).

McCarthy, M. M. (1981), 'Church and State: Separation or Accommodation?' *Harvard Educational Review*, 51/3: 373–94.

MacCurtain, M. (1972), *Tudor and Stuart Ireland* (Dublin: Gill and Macmillan).

—— (1974), 'Pre-Famine Peasantry in Ireland: Definition and Theme', *Irish University Review*, 4: 188–98.

McDonagh, E. (1972), 'Discriminatory Provisions of the Irish Constitution', *Doctrine and Life*, 22: 320–32.

—— (1985), 'Northern Ireland and Irish Responsibility', *Furrow*, 36/6: 345–54.

MacDonagh, O. (1975), 'The Politicization of the Irish Catholic Bishops', *Historical Journal*, 18: 37–53.

—— (1977), *Ireland: The Union and its Aftermath* (London: Allen & Unwin).

—— (1983), *States of Mind: A Study of Anglo-Irish Conflict, 1780–1980* (London: Allen & Unwin).

MacDonald, M. (1986), *Children of Wrath: Political Violence in Northern Ireland* (Cambridge: Polity Press).

McFarlane, W. G. (1979), 'Mixed Marriage In Ballycuan, Northern Ireland', *Journal of Marriage and the Family*, 10: 191–205.

Mác Gréil, M. (1977), *Prejudice and Tolerance in Ireland* (Dublin: College of Industrial Relations).

—— and O'Gliasain, M. O. (1974), 'Church Attendance and Religious Practice of Dublin Adults', *Social Studies*, 3/2: 163–212.

MacIver, M. A. (1989), 'A Clash of Symbolism in Northern Ireland: Extremist and Moderate Political Élites', *Review of Religious Research* (forthcoming).

McKee, E. (1986), 'Church–State Relations and the Development of Irish Health Policy: The Mother-and-Child Scheme 1944–53', *Irish Historical Studies*, 25: 159–94.

McKeown, M. (1977), 'Integrated Education: The Debate Surveyed', *Furrow*, 28/1: 4–17.

McKernan, J. (1980), 'Pupil Values as Social Indicators of Intergroup Differences in Northern Ireland', Harbison and Harbison 1980: 128–40.

McNamara, K. (1972), 'Church and State', *Doctrine and Life*, 22: 339–50.

McRedmond, L. (1974), 'The Normality of the Irish Catholic', *Month*, 7/8: 672–5.

—— (1977), 'The Tyranny of the Majority', *Tablet*, 19 Mar., 269–72.

—— (1985), 'A Brawling Church: The Malaise of Irish Catholicism', *Doctrine and Life*, 34/7.

Maginn, M. (1986), 'The Zigs and Zags of Irish Catechetics', *Furrow*, 37/9: 574–80.

Manhattan, A. (1971), *Religious Terror in Ireland* (London: Paravision Books).

Manning, M. (1970), *The Blueshirts* (Dublin: Gill and Macmillan).

Mansergh, N. (1974), 'The Government of Ireland Act, 1920: Its Origins and Purposes. The Workings of the "Official" Mind', in Barry 1974.

—— (1975), *The Irish Question 1840–1921*, 3rd edn. (Toronto: Univ. of Toronto).

Mark, R. (1979), 'Inter-denominational Education: What Type of Movement? Theological, Sociological and Ecumenical Reflections on the All Children Together Movement', short MA diss., Univ. of Hull/Irish School of Ecumenics, Dublin.

Martin, D. (1978), *A General Theory of Secularization* (Oxford: Blackwell).

Marx, K. (1971), *The Early Texts*, ed. D. McLellan (Oxford: Blackwell).

Maynooth (1960), *Acta et Decreta Concilii Plenarii Quod Habitum est apud Maynooth, 7–15 August 1956* (Dublin: Gill and Sons).

Mehl, R. (1970), *Sociology of Protestantism* (London: SCM).

Methodist Church in Ireland (1983), *Minutes of Conference* (Eniskillen: Methodist Church in Ireland).

Miles, R. (1989), *Racism* (London: Routledge).

Miller, D. W. (1973), *Church, State and Nation in Ireland 1891–1921* (Dublin: Gill and Macmillan).

—— (1975), 'Irish Catholicism and the Great Famine', *Journal of Social History*, 9: 81–98.

—— (1978a), *Queen's Rebels: Ulster Loyalism in Historical Perspective* (Dublin: Gill and Macmillan).

—— (1978b), 'Presbyterianism and "Modernization" in Ulster', *Past and Present*, 80: 66–90.
MOORE, B. (1973), 'Race Relations in the Six Counties: Colonialism, Industrialisation and Stratification in Ireland', *Race XIII*, 14/1: 21–42.
MORGAN, A. and PURDIE, B. (eds.) (1980), *Ireland: Divided Nation, Divided Class* (London: Ink Press).
MOUFFE, C. (1979), 'Hegemony and Ideology in Gramsci', in C. Mouffe (ed.), *Gramsci and Marxist Theory* (London: Routledge & Kegan Paul).
MOXON-BROWNE, E. (1983), *Nation, Class and Creed in Northern Ireland* (Aldershot: Gower Press).
—— and IRVINE, C. (1989), 'Not many Floating Voters Here', *Fortnight*, 273: 7–9.
MURPHY, C. (1980), *School Report* (Dublin: Ward River Press).
MURPHY, J. (1959), The Religious Problem in English Education: the Crucial Experiment (Liverpool: Liverpool University Press).
—— (1984), 'Censorship and the Moral Community in Ireland', in B. Farrell (ed.), *Communication and Community in Ireland* (Cork and Dublin: Mercier), 51–64.
MURPHY, J. H. (1984), 'The Role of Vincentian Parish Missions in the "Irish Counter-Reformation" of the Mid-Nineteenth Century', *Irish Historical Studies*, 24/94: 152–71.
MURRAY, D. (1983), 'Schools and Conflict', Darby 1983: 136–50.
—— (1985), *Worlds Apart: Segregated Schools in Northern Ireland* (Belfast: Appletree Press).
NELSON, S. (1984), *Ulster's Uncertain Defenders: Loyalists and the Northern Ireland Conflict* (Belfast: Appletree).
New Ireland Forum (1983), *A Comparative Description of the Economic Structure and Situation, North and South* (Dublin: Stationery Office).
—— (1983–4), *Report*, i, and *Reports and Proceedings*, ii–xiii (Dublin: Stationery Office).
New University of Ulster (1978), *Segregation in Education* (Derry: Institute of Continuing Education).
NIC GHIOLLA PHADRAIG, M. (1978), 'The Legitimation of Social and Ultimate Values in the Context of Religious Norms: A Study of Social and Religious Factors', Ph.D. Thesis, Univ. College, Dublin.
—— (1981), 'Alternative Models to Moral Reasoning', CISR, Lausanne, 16th Annual Conference, 364–9.
NOTTINGHAM, E. (1954), *Religion and Society* (New York: Random House).
—— (1972), *Religion: A Sociological View* (New York: Random House).
O'BRIEN, CONOR CRUISE (1957), *Parnell and his Party, 1880–1890* (Oxford: OUP).
—— (1974), *States of Ireland* (St Albans: Panther Books).

O'Carroll, J. P. (1983), 'Some Sociological Aspects of the 1983 Abortion Referendum Debate in the Republic of Ireland', Paper to Sociological Association of Ireland Annual Conference, Belfast.

O'Connell, D. (1846), *The Select Speeches of Daniel O'Connell*, ed. John O'Connell (Dublin: Gill).

O'Connell, D. (1982), 'Sociological Theory and Irish Political Research', Kelly, *et al.* 1982: 186–98.

O'Connor Lysaght, D. R. (1980), 'British Imperialism in Ireland', Morgan and Purdie 1980: 12–32.

O'Doherty, H. (1980), 'The Dalkey School Project', short MA diss., Univ. of Hull/Irish School of Ecumenics, Dublin.

O'Donnell, E. E. (1976), *Northern Ireland Stereotypes* (Dublin: College of Industrial Relations).

O'Dowd, L., Rolston, B., and Tomlinson, M. (1980), *Northern Ireland: Between Civil Rights and Civil War* (London: CSE Books).

OECD (1965), *Investment in Education* (Dublin: Stationery Office).

—— (1969), *Reviews of National Policies for Education: Ireland* (Paris: Organization for Economic Cooperation and Development).

O'Grada, C. and Boyle, P. (1984), 'Fertility Trends, Excess Mortality and the Great Irish Famine', Paper given at a meeting of the British Society for Population Studies, LSE, 6 July.

O'Halloran, C. (1987), *Partition and the Limits of Irish Nationalism: An Ideology under Stress* (Dublin: Gill and Macmillan).

O'Malley, P. (1983), *The Uncivil Wars* (Belfast: Blackstaff).

One in Christ (1975), 'Joint Church Recommendations for the Preparing of Interchurch Couples for Marriages', *One in Christ* (1975), 385–97.

Open University (1973), *Education in Great Britain and Ireland: A Source Book*, eds. R. Bell, G. Fowler, and K. Little (London and Boston, Mass.: Routledge & Kegan Paul).

O'Shea, J. (1986), *Priests, Politics and Society in Post Famine Ireland: A Study of Co. Tipperary 1850–91* (Dublin: Wolfhound Press).

Ottaviani, A. (1953), *Doveri dello Stato cattolico verso la religione* (Rome: Libreria del P. Ateneo Lateranense).

O'Tuathaigh, G. (1986), Lecture on 'People in Ireland', BBC Radio Ulster, as reported by *Irish Times*, 2 Dec.

Paisley, I. (1966), *Messages from a Prison Cell* (Belfast: Puritan Publishing Company).

—— (1970), *Northern Ireland: What is the Real Situation?* (Greenville, SC: Bob Jones Univ. Press).

Papal Encyclicals (1939), *Selected Papal Encyclicals and Letters*, i, *1896–1931* (London: Catholic Truth Society).

Patterson, H. (1980), *Class Conflict and Sectarianism: The Protestant*

Working Class and the Belfast Labour Movement 1868–1920 (Belfast: Blackstaff Press).
PAUL VI (1970), 'Matrimonia Mixta', *Furrow*, 21: 793–8.
PEARSE, P. H. (1980), *The Letters of P. H. Pearse* (Gerrards Cross: Colin Smythe).
PEILLON, M. (1982), *Contemporary Irish Society: An Introduction* (Dublin: Gill and Macmillan).
PHILBIN, Bp. W. (1975), 'Catholic Education', *Sunday Press*, 16 Feb., 2.
—— (1976), 'Schools and Integrated Education', interview, RTE, 3 Oct.
PIUS XI (1929), *Christian Education of Youth (Divini Illius Magistri)* (London: Catholic Truth Society).
—— (1930), *On Christian Marriage (Casti Conubii)* (London: Catholic Truth Society).
—— (1931), *On Reconstructing the Social Order (Quadragesimo Anno)* (London: Catholic Truth Society/Oxford: Catholic Social Guild).
PORTELLI, H. (1974, 1976), *Gramsci et la question religieuse* (Paris: Anthropos, 1974); Italian trans., with texts from the critical edn. of the *Quaderni*: *Gramsci et la questione religiosa* (Milan: Gabriele Mazotta, 1976).
PRAGER, J. (1986), *Building Democracy in Ireland* (Cambridge: CUP).
Presbyterian Church in Ireland (1895–1986), *Minutes of the General Assembly* (Belfast: Church House).
—— (1930–86), *Annual Reports* (Belfast: Church House).
PRINGLE, D. G. (1985), *One Island, Two Nations? A Political Geographical Analysis of the National Conflict in Ireland* (Letchworth, GB: Research Studies Press and New York: Wiley).
PROBERT, B. (1978), *Beyond Orange and Green: The Political Economy of the Northern Ireland Crisis* (London: Zed Press).
Protestant and Catholic Encounter, *Pace* (1969–), Belfast.
REA, D. (ed.) (1982), *Political Co-operation in Divided Societies* (Dublin: Gill and Macmillan).
REED, B. (1984), *Ireland: The Key to the British Solution* (London: Larkin Publications).
REID, J. S. (1853), *History of the Presbyterian Church in Ireland*, i and ii, *1603–1700* (London: Whittaker & Co.).
ROCCA, T. LA (1981), *Gramsci e la Religione* (Brescia: Queriniana).
ROCHE, D. J. D., BIRRELL, W. D., and GREER, J. I. (1975), 'A Socio-Political Opinion Profile of Clergymen in Northern Ireland', *Social Studies*, 4/2: 143–51.
ROLSTON, B. (1983), 'Reformism and Sectarianism: The State of the Union after Civil Rights', Darby 1983: 197–224.
—— and TOMLINSON, M. (1982), 'Spectators at the "Carnival of Reaction?"

Analysing Political Crime in Ireland', Kelly, O'Dowd, and Wickham, 1982: 21–43.
ROSE, R. (1971), *Governing without Consensus: An Irish Perspective* (London: Faber & Faber).
—— (1976), *Northern Ireland: A Time of Choice* (London: Macmillan).
RUMPF, E. and HEPBURN, A. (1977), *Nationalism and Socialism in Twentieth Century Ireland* (Liverpool: Liverpool Univ. Press).
RUSSELL, J. L. (1972), *Civic Education in Northern Ireland* (Belfast: Northern Ireland Community Relations Commission).
RYAN, L. (1979), 'Church and Politics: The Last Twenty-five Years', *Furrow*, 30/1: 3–18.
SALTERS, J. (1970), 'Attitudes towards Society in Protestant and R.C. School Children in Belfast', MA diss., Queen's University Belfast.
SENIOR, H. (1966), *Orangeism in Ireland and Britain, 1795–1836* (London: Routledge & Kegan Paul).
SHIPMAN, M. D. (1972), *Childhood: A Sociological Perspective* (Slough: NFER).
SMITH, A. B. (1980), 'Interdenominational Religious Education in Africa: The Emergence of Common Syllabuses', MA thesis, Univ. of Hull/Irish School of Ecumenics, Dublin.
SMITH, A. D. (1971), *Theories of Nationalism* (London: Duckworth).
—— (1976), 'The Formation of Nationalist Movements' in Smith (ed.), *Nationalist Movements* (London: Macmillan).
—— (1979), *Nationalism in the Twentieth Century* (Oxford: Martin Robertson).
—— (1981), *The Ethnic Revival* (Cambridge: CUP).
—— (1983), 'Nationalism and Classical Social Theory', *British Journal of Sociology*, 34: 19–38.
SMYTH, C. (1987), *Ian Paisley: Voice of Protestant Ulster* (Edinburgh: Scottish Academic Press).
SOUTHERN, R. W. (1970), *Western Society and the Church in the Middle Ages* (New York: Penguin).
SPENCER, A. E. C. W. (1968), 'An Evaluation of Roman Catholic Educational Policy in England and Wales 1900–60', Jebb 1968: 165–221.
—— (1973), 'Christian Proposals for the Irish Churches', *Month*, Jan.
STEWART, A. T. Q. (1977), *The Narrow Ground: Aspects of Ulster 1609–1969* (London: Faber & Faber).
Tablet, The (1983), Articles on Anti-Abortion Amendment to the Irish Constitution: 346–7, 816–18, 832, 887, 904.
—— (1983, 1984), Articles on the New Ireland Forum (1983), 264–5; (1984), 52–3, 60–1.
TAYLOR, P. (1983), 'The Lord's Battle: An Ethnographic and Social Study

of Paisleyism in Northern Ireland', Ph.D. thesis, Queen's University, Belfast.

THOMPSON, J. (1981), 'The Westminster Confession', Haire, et al., 1981: 4–27.

TITLEY, E. B. (1983), *Church, State and the Control of Schooling in Ireland 1900–44* (Dublin: Gill and Macmillan).

TOBIN, C. (ed.) (1985), *Seeing is Believing: Moving Statues in Ireland* (Mountrath: Pilgrim Press).

TODD, J. (1987), 'Two Traditions in Unionist Political Culture', *Irish Political Studies*, 2: 1–26.

TOWNSHEND, C. (1987), 'The Partitioned Mind', *Irish Literary Supplement*, 6/2: 12.

TURNER, E. B., TURNER, I. F., and REID, A. (1980), 'Religious Attitudes in Two Types of Urban Secondary School: A Decade of Change?' *Irish Journal of Education*, 14: 43–52.

VAN NOORT, G. (1960), *Tractatus De Ecclesia Christi*, 5th edn. (Hilversum: Paulus Brand).

Vatican II (1967), *Documenti del Concilio Vaticano II*, 5th edizioni in Italian and original Latin (Bologna: edizioni Dehoniane).

VERMEERSCH, A. S. J. (1904), *Quaestiones De Iustitia* (Bruges: Beyaert Publishers).

VOS, J. G. (1980), *The Scottish Covenanters* (Pittsburgh, Pa.: Crown & Covenant Publications).

WALLIS, R. and BRUCE, S. (1985), 'Sketch for a Theory of Conservative Politics', *Social Compass*, 32: 145–62.

——— (1986), *Sociological Theory, Religion and Collective Action* (Belfast: Queen's Univ.).

——— and TAYLOR, D. (1986), *'No Surrender': Paisleyism and the Politics of Ethnic Identity in Northern Ireland* (Belfast: Queen's Univ. Social Studies Dept.).

WALSH, B. (1970), *Religion and Demographic Behaviour in Ireland* (Dublin, Economic and Social Research Institute, Paper No. 55).

—— (1975), 'Trends in the Religious Composition of the Population in the Republic of Ireland, 1946–71', *Economic and Social Review*, 6: 543–56.

WATT, D. (ed.) (1981), *The Constitution of Northern Ireland: Problems and Prospects* (London: Heinemann).

WEBER, M. (1930), *The Protestant Ethic and the Spirit of Capitalism* (London: Allen & Unwin).

—— (1964), *The Theory of Social and Economic Organization* (New York: Free Press of Glencoe).

—— (1965), *The Sociology of Religion* (London: Methuen).

WHITE, J. (1975), *Minority Report: The Protestant Community in the Irish Republic* (Dublin: Gill and Macmillan).

WHYTE, J. H. (1960), 'The Influence of the Catholic Clergy on Elections in Nineteenth Century Ireland', *English Historical Review*, 75.

—— (1978), 'Interpretations of the Northern Ireland Problem: An Appraisal', *Economic and Social Review*, 9/4: 257–82.

—— (1980), *Church and State in Modern Ireland 1923–79* (Dublin: Gill and Macmillan).

—— (1981), 'Religion and Conflict in Northern Ireland', Conference Internationale de Sociologie de Religion, 372–6.

—— (1983), 'How much Discrimination was there under the Unionist Regime, 1921–68?' Gallagher and O'Connell, 1983: 1–35.

—— (1986a), 'Interpretations of the Northern Ireland Problem', British Association for Irish Studies Conference, Univ. of Keele, Apr.

—— (1986b), 'Recent Developments in Church–State Relations', *Seirbhis Phoibli*, 6/3: 4–10.

WRIGHT, F. (1973), 'Protestant Ideology and Politics in Ulster', *European Journal of Sociology*, 14/2: 213–80.

—— (1987), *Northern Ireland: A Comparative Analysis* (Dublin: Gill and Macmillan).

Index

abortion debate and referendum 155–8
  anti-abortion movement 133
  and protestant churches 156–7
  see also Pro-Life Amendment Campaign; Society for the Protection of the Unborn Child
ACT, *see* All Children Together
Act of Union (1800) 52, 53
Adoption Act (1974) 154
Adoption Bill (1952) 146–7
All Children Together 181
Alliance Party 95, 97
Anglo-Irish:
  before 1922, *see* ascendancy
  from 1922, *see* protestants, Southern
Anglo-Irish Agreement (Hillsborough Accord 1985) 95, 97, 107, 108
Anglo-Norman 25–6, 68
anti-abortion movement, *see* abortion
Apprentice Boys, *see* religious political societies; Siege of Derry
Armistice Day 91
Armstrong, Revd David 119–20
ascendancy 36, 37–8, 42, 47–8, 61–2
  decline of 52, 53
  development of 29–33
  and dissenters (presbyterians) 29–31, 33, 36
  Northern 53–4
  parliament 32, 36, 40; *see also* Grattan's parliament
  and Roman catholics 31, 33, 36
  successor to, *see* protestant–loyalist bloc

'B' specials 95–7
bad sectarianism and violence 13–17
banding tradition 29, 40–1, 58, 66
Barkley, J. M. 28, 45, 214
Belfast shipworkers and riots 62–4
Belfast workers' demonstration (1932) 93
Bew, Gibbon, and Patterson 4, 95
BICO, *see* British and Irish Communist Organization
Binchy, William 162
Black Oath 30
Black Preceptory, *see* religious political societies

'Bloody Sunday' 127, 131–2
Boyce, D. G. 67–8, 77
Breslin and Weafer 9–10, 11
British identity of loyalists, *see* protestant–loyalist bloc
British and Irish Communist Organization 106
British government:
  in Northern Ireland 95, 97
  power and policies in Ireland (to 1800) 36, 47–8, 52
British historical bloc 26
Browne, Dr Noel 148–9
Buckley, A. D. 99, 123–4

Calvinism 28
  see also presbyterian church
Cantwell, Bishop 82–3
Casey, Charles 146
Catholic Association 69
catholic church, *see* Roman catholic church
catholic emancipation 69, 71
Catholic Herald and Standard 167–8
catholic Ireland, ideology of 90–1
catholic nationalism 93, 131–2
  see also catholic–nationalist bloc, dominant beliefs
catholic–nationalist bloc 17–18, 22–4, 56, 61–2
  attitudes to violence 13–16, 80
  composition and class formation 23, 69, 70–1, 73, 76–80, 89, 92, 93
  contradictions 85, 151–2
  dominant beliefs and ideology 23–4, 100, 103, 104–6, 108–17, 127–30, 131–2
  migration 91, 92
  politics in North 94
  and protestant–loyalist bloc 76, 98–9
  subordination in the North 92, 95–6, 97–8
  see also cultural similarities; Irish Party; violence
catholic nationalists:
  in the North 92–4, 97–8, 99
  and Roman catholic church 76, 80–5, 93, 170

catholic nationalists (*cont.*):
  and state law 133
  see also Roman catholics
catholic–protestant opposition 35, 66–7
catholic schools, *see* schooling
censorship 145
Charles I 28, 29, 31
Charles II 31
Christian Brothers:
  and nationalism 110
  and schooling 74, 110, 171–2
church attendance, *see* protestants; Roman catholics
Church of Ireland:
  and abortion 156, 157
  disestablishment of (1869) 76
  and divorce 167
  and education 171, 177, 178–9, 184
  founding of 28
  order (organization) of 101–2, 119
  and presbyterians 28–33
  in the Republic 90
  see also ascendancy
church–state, *see* Roman catholic church
civil and human rights issues 17, 94, 135–7, 215
civil society 21–2, 125–6, 152
class and class structure:
  in 17th century 27–8, 29–30, 32, 33
  in 18th century 37–40, 42–4, 45, 47
  in 19th century 54, 66, 68–71, 76
  in 20th century (North) 95, 97–8, 112
  in 20th century (South) 90–1, 110–11, 112, 150
  and presbyterian subordination 33, 35, 36–7
  subordination of Ulster catholics within 95–8
  suppression of issues of 66, 110–12
  see also catholic–nationalist bloc; protestant–loyalist bloc
Compton and Coward 198–200, 201
Compton and Power 92
Communists, *see* British and Irish Communist Organization
Congregation of Propaganda 172
Connolly, James 112
contraception 154
Cooke, Henry 60
covenant:
  and loyalism, Orangeism 54
  politics 31, 58, 107, 121, 228

and social contract theory 41, 122
Solemn League and (1639) 28–9; (1911) 86; (1971) 121
and violence 130–1
see also protestant–loyalist bloc, dominant beliefs
Counter-Reformation 27
see also Tridentine
Cox, W. H. 131–2
Cromwell, Oliver 3, 27, 31
Cullen, Cardinal James 75, 76, 77, 80, 83–4, 113, 128
cultural nationalism 109–10
cultural similarities 98–9
culture, *see* catholic–nationalist bloc; Irish language; protestant–loyalist bloc; religion, of the laity; state culture
Cushnahan, John 168

Dalkey School Project 181–2
Daly, Bishop Cathal 160, 169, 184–5
Daly, Bishop Edward 180
Davis, Thomas 80
De Tocqueville, A. 81, 134
De Valera, Éamon 80, 90–1, 100, 110, 135, 139, 140–1, 143, 149
Defenders 44, 48, 50
Democratic Unionist Party 15–16, 107
  support for 107–8
  see also fundamentalism; Paisley, Ian
Derry, *see* Siege of Derry
Dingle mission (1846) 73–5
discrimination in Ulster, *see* class structure, subordination
dissenters, *see* presbyterians
divorce:
  and constitution 137
  debate and constitutional referendum 160–70
  in Ireland 145, 160
  and protestant churches 162, 167
  and Roman catholic bishops 162–3, 164, 167–8, 169–70
dominant beliefs, *see* catholic–nationalist bloc; protestant–loyalist bloc
Dominian, Dr Jack 167
Dukes, Alan 164
DUP, *see* Democratic Unionist Party

Easter (Dublin) rising 8, 87, 129
economies of Ireland 29–30, 33–4, 38–40, 42–4, 57–8, 68–71, 91, 153
ecumenism 12, 106

Index

education in Ireland, *see* Church of Ireland; integrated schooling; Roman catholic church; schooling; schools in Ireland
elective affinity 61
Elizabeth I 26
endogamy and mixed marriage, *see* mixed marriage
evangelical revival 28, 59–62, 114
Eversley, D. 97

famine 70, 72
Fenians 80, 128
Fianna Fáil 100, 109, 129, 135, 147, 158–9, 162
Fine Gael 109, 155, 162
Firhouse survey on schooling 182–3
FitzGerald, Garret 8, 17, 155, 161–2, 169
free presbyterians 13–14, 102, 118
  and violence 13
  see also Democratic Unionist Party
fundamentalism 102, 121, 227
  and violence 13
  see also free presbyterians

Gaelic Athletic Association 109–10
Gaelic revival, *see* Irish language
Gaels 67–8
Gaine, M. 191
Gellner, *see* industrialization, theory
gerrymandering 95
Gibbon, P. 43–4, 48–9, 55, 57, 59
Gladstone, William E. 64, 85
good religiosity 9–12, 16, 128
Government of Ireland Act (1920) 87
Gramsci, Antonio 4–5, 18–22, 84–5, 130, 142, 172
Grattan, Henry 40
Grattan's parliament 42, 47–8, 50, 52
Greeley, Andrew 186–7

Haughey, Charles 159, 165
hegemonic culture 143
hegemonic structure of state 150
hegemony 4, 13, 21, 22, 78, 84–5, 90, 118, 126, 127–8, 170
  see also catholic–nationalist bloc; protestant–loyalist bloc
Henry VIII 25–6
Hickey, J. 7–8, 121
Hillsborough, *see* Anglo-Irish Agreement
'historic integrity', *see* Ireland
historical bloc 4, 18, 19–20
  see also catholic–nationalist bloc; protestant–loyalist bloc
home rule 64–6, 84, 85–6, 118
Huguenots 36
human rights, *see* civil and human rights issues
Hume, John 105
hunger strikes 131
Hurley, Michael 101
Hyde, Douglas 79

Ignatius of Loyola 113
industrialization:
  theory of 57–8
  of Ulster 38–9, 58
industry North and South 98
integrated schooling (education):
  in Africa 193
  movement 180–3
INTO, *see* Irish National Teachers' Organization
Ireland:
  'historic integrity' of 109, 158–9
  as a unit of analysis 4
Irish civil war (1922–3) 80, 92
Irish Constitution (Republic of Ireland 1937):
  and church teaching 133, 137, 141–2, 153–4
  and national territory 139
  preamble 135–6
  and senate 138
  see also abortion debate and referendum; divorce; papal teaching and Irish constitution; religion, and law
Irish Free State 4, 89, 91, 92
  Constitution of 135, 136–7
Irish Labour Party 110, 145, 162
Irish language:
  and Gaelic revival 78–9
  and the State 109–10, 139
Irish medieval church 25–6
Irish mobility survey 7
Irish nation and nationalism 8, 42, 51, 52, 67–8, 77–80, 105–6, 139
  see also catholic–nationalist bloc
Irish National Teachers' Organization 173
Irish nationalists, *see* catholic–nationalist bloc
Irish parliament 32
  see also ascendancy, parliament; Grattan's parliament

Index

Irish Party 84–5, 87
  and Parnell 80–1, 84
  and the Roman catholic church 84, 174–5
Irish Volunteers 86–7
island-wide sense of identity 26
Ius Publicum 140

Jackson, J. 7
Jacobite rising (1715) 37
James I 27, 28
James II 31, 32, 35, 40, 123
Jesuits, *see* Roman catholic, religious orders
John-Paul II, Pope 154–5, 161

labour party, *see* Irish Labour Party
land question 55, 80, 83
Larkin, E. 4, 70, 71, 72
Larkin, James 87, 112
Latimer, W. T. 44
Lee, R. 198, 199, 200, 201, 213
Lemass, Sean 131, 153
Leo XIII 83, 136, 138, 142–3
Lévi-Strauss, C. 122, 124
liberal party and the liberals 56, 81, 174
liberal theology, *see* presbyterian church; Roman catholic church
librarian of Co. Mayo, case of 143–4
Londonderry, *see* Siege of Derry
loyalism, ambiguity of, *see* protestant–loyalist bloc
Lundy, Robert 123

McAllister, I. 7–8, 94
MácGréil, M. 15, 93, 128
McCann case (1910) 212–15
McGee case (1971–3) 154
McHale, Archbishop 128, 172
McKernan, J. 99
McNamara, Archbishop 155, 162–3, 166
McQuaid, Archbishop 117, 140, 147, 148
McRedmond, Louis 168
majority, concept of 77, 108, 136
  see also monopoly of the majority
Mallon, Seamus 105
marching tradition, *see* protestant–loyalist bloc
Maynooth 48, 99
methodist church 157
methodists 199–200

migration, *see* catholic–nationalist bloc
Miller, D. W. 28–9, 38, 43, 45–6, 54–5, 57, 58, 59, 60, 72–3, 107, 121
mixed marriage 6, 90, 198–226
  in Ireland 201–4, 209–11, 215–16, 221–6
  and the McCann case 212–15
  and *Ne Temere* 211–12
  and popular culture 203, 225
  and protestant churches 202, 204, 223, 225
  Roman catholic history of 204–12
  and the Tilson case 216
  and Vatican II 216–19
Molyneux, William 41
modernization, *see* industrialization
monopoly catholicism, *see* Roman catholic church
monopoly of the majority 108
Moran, D. P. 78–9
multi-denominational education, *see* integrated schooling
Murray, Archbishop 172
Murray, D. 188
myths:
  of protestant loyalists 122–5
  of Roman catholics 124–5

nation 57–8
  see also catholic–nationalist bloc; Irish nation and nationalism
nationalism, secular 130
nationalists, *see* nation
national–popular consciousness 100, 133
natural law and the common good 133, 134, 141, 159–60, 169–70
New Ireland Forum 158–60
non-subscribers, *see* presbyterians
Northern catholics, *see* Roman catholics, in the North
Northern Ireland:
  forming of 85–8, 95, 121
Northern Ireland fertility survey (1983) 199
Northern Ireland Labour Party 94, 112
Northern Star 46–7, 50
Nottingham, E. 72

O'Brien, Conor Cruise 8
O'Carroll, J. P. 116
O'Connell, Daniel 69, 77–8, 81, 85, 106
O'Dowd, Rolston and Tomlinson 4, 96

Index

Official Sinn Fein, *see* Workers' Party
O'Fiaich, Cardinal 8
O'Halloran, Clare 105
O'Neill, Phelim 29
O'Neill, Terence 97
Orangeism and Orange Order 16, 44, 48–50, 53, 54–5, 56–7, 58, 61, 64, 65, 66–7, 101–2
   see also religious political societies

Paisley, Ian 8, 101, 120–1
   and violence 130
Pan-Celtic General Alliance, *see* Irish language
pan-protestantism 117–18
papal teaching and Irish constitution 136–8, 140, 142–3
Parnell, Charles S. 80–1, 83
partition of Ireland 87–8
Patriot Party 38, 40
Pearse, Padraig 129
people of Ireland 158
   see also Irish nation and nationalism
people of Ulster 120
Philbin, Bishop 101, 180–1
pillarization 99–100
PIRA, *see* Provisionals
plantations of Ulster 27
political party system 92, 93–5, 109–10
political religion 22, 133, 171
political–religious mediation 133
Popery Act (1704) 36
popular religion, *see* religion, of the laity
Poynings Law (1494) 36
presbyterian church:
   in Ireland 30–1, 32, 34
   order (organization) 102, 119
   conservative, liberal, and revolutionary tendencies 45–7, 55–6, 59, 65, 119
   non-subscribing 34
   seceders 34–5
   settlers 28–9
presbyterians (dissenters) 5, 31, 32, 35, 36–7, 45, 55, 119–21
   and Calvinism 120, 125
   and class, *see* class and class structure
   and education 171
   and Roman catholics 50–1, 55
   see also abortion debate and referendum; covenant, politics; Church of Ireland; divorce, debate; mixed marriage

Progressive Democrats 111
Pro-Life Amendment Campaign 116, 155
Propaganda Fidei 82, 172
proselytism 60, 73–4, 75
protestant:
   anti-catholicism 33, 66–7
   bloc 89
   British identity 58
   and catholic opposition 35
   churches 101–2
   classes 33–4, 118
   divisions in Ireland 118–119
   fundamentalism 13, 118, 120–1
   Irish, *see* ascendancy
   liberty 117–18, 121
   marching tradition 13, 66
   planter attitudes 27
   proselytism 73–5
   theology 34–5, 61, 119; *see also Westminster Confession*
   workers of Belfast and riots 62–4
   see also myths; schools in Ireland
protestant churches and 'the Troubles' 8
protestant–loyalist bloc 17–18, 22–4, 53
   attitudes to clergy 121
   and class formation 64–7, 89, 97–8
   composition (alliance) 22, 93
   and British alliance 95, 97, 228
   and British identity 108, 117
   class politics 112
   dominant beliefs 23–4, 103–4, 106–8, 117–25, 130–1, 228–9
   and ethnicity 106
   identity contradictions 107–8, 228
   monopoly tendencies 120–1
   and nation 106; *see also* Irish nation
   oppositional culture and politics 122, 227–9
   origins and conditions for growth 54
   political base 102
   politics 94–7
   and popular religion and culture 56–64, 120, 121
   schools 176–7
   and the state 89–90, 92–9
   suppression of class conflict 93, 112
   see also covenant; Orangeism and Orange Order; religious political societies
protestant loyalists 89

protestantism:
  popular 56–62, 62–4, 66–7, 120–1
  see also myths; pan-protestantism; protestant–loyalist bloc
protestants 33, 34–5
  religious practice of 11–12
  Southern 89, 90
  and violence 130–1; see also Ulster Clubs
  see also ascendancy; Church of Ireland; free presbyterians; fundamentalism; presbyterians; protestant–loyalist bloc
Provisional Sinn Fein 13–15, 92, 94, 112, 129
Provisionals (Provisional Irish Republican Army) 127–8, 129–30

race and racism 117–18
rebellion (1798), see United Irishmen and rebellion
Redmond, John 87
Reformation in Ireland 25–7
regium donum 31, 34, 46
Relief Act:
  (1778) 40
  (1794) 70
religion:
  and the ethos of the Irish constitution (1937) 134–8
  of the laity (people) 116, 130, 133; see also catholic–nationalist bloc; myths; protestant–loyalist bloc; protestantism, popular; Roman catholics
  and law in the Republic 133–70
  and loyalism, see protestant–loyalist bloc
  and nationalism 68; see also catholic–nationalist bloc
  and 'the Troubles', violence 7–9, 13–16, 128–30, 131; see also catholic–nationalist bloc; protestant–loyalist bloc
religious political societies 102, 123
republicanism 103, 111–12
  see also catholic–nationalist bloc
republicans 79–80, 87–8
reunification 131
revival, see evangelical revival
riots 37, 62–4
Robinson, Peter 108
Roman catholic:
  beliefs and popular culture 71–6, 113–17
  bishops and clergy 40, 47, 71–2, 75–6, 80–2, 99, 101, 113–15, 117, 124, 147–50, 152–3, 156, 159
  lay opposition to clergy 194–5
  liberals 77, 101, 153
  monopoly (monopoly catholicism) 122, 132, 151, 155, 157, 170, 185, 191, 227, 299–30
  political forms 229
  religious orders 101, 113–15
  schools, see schooling; schools in Ireland
  sexual and family mores (rural) 76–7, 117, 144–5
  social teaching and Irish constitution 137–8, 139, 140–2; see also Adoption Act; Adoption Bill
  theology of church and churches 8, 110, 150, 196–7, 216–19
  traditional spirituality 113–17
Roman catholic church:
  and authority 77, 115–16, 152–3, 185, 191, 193
  and British state 40, 83–4, 134
  and education, see schooling; schools in Ireland
  as legitimator of Southern state 152
  and Mother and Child Scheme 147–9
  organization 99–100
  and papacy 82–3
  political power 17, 149
  and referendum (1972) 153
  and secularism, secularization 185, 191–2, 197
  and socialism 111, 145
  and state 100
  and subsidiarity 142, 151
  see also abortion debate and referendum; Adoption Act; Adoption Bill; divorce, and constitution; education in Ireland; integrated schooling; Irish nation; mixed marriage; religion, and law
Roman catholics:
  and authoritarianism 13
  in the North 97–8, 99, 104–5; see also class and class structure
  religious practice and commitment of 9–12, 72–3

Index

and violence 13–15, 79–80, 127–31
  see also catholic–nationalist bloc
Rose, R. 105, 187–8
Royal Black Preceptory, see religious political societies
Rumpf and Hepburn 111
Russell, J. L. 188

sacramental test, see Test Act (sacramental)
Salters, J. 188
schooling:
  contemporary controversy over 184–9
  Roman catholic ideology and theology of 189–97
schools in Ireland:
  clergy control of 172–3
  history of 171–5
  'model' 173
  Northern 99–100, 176–7
  segregated 179–80
  Southern 177–80
Scottish Kirk 28–9
SDLP, see Social Democratic and Labour Party
seceders, see presbyterians
sectarianism, see bad sectarianism and violence
secularism, fear of, see Roman catholic church
secularization 9–11
  see also Roman catholic church
Senior, H. 41, 48, 50
Seven Years War (1756–63) 40
Siege of Derry (1688–9) 32, 123–4
Sinn Fein 80, 87–8, 92
  see also Provisional Sinn Fein; Workers' Party
Smyth, Martin 15
social contract theory 38, 41
  see also covenant
Social Democratic and Labour Party 92–3, 94, 105, 111, 158
socialism 111–12
  see also Roman catholic church
Society for the Protection of the Unborn Child 157–8
sociologie religieuse 9, 16, 143
Solemn League and Covenant, see covenant
Spencer, A. 185–6
SPUC, see Society for the Protection of the Unborn Child

Starkie, Dr William 174
state culture 141–2
state and violence 126–7
Statistical Survey of Ireland 38–9
Statutes of Kilkenny 26
Steelboys 43
Stormont parliament 89, 94, 95, 96
subsidiarity, see Roman catholic church
surveys, use of 187

Test Act (sacramental) 36–7
Tilson case (1951–2) 216
Trail, Revd William 29
Treaty of Limerick 32
Tone, Wolfe 47, 51
Trent, Council of, see Tridentine
Tridentine:
  decree on marriage 205–6
  religion 73–5, 84, 229
'the Troubles' 94–5
  interpretations of 1–2
  and religion 1, 3

Ulster Clubs 102
Ulster Conflict, see 'the Troubles'
Ulster protestants, see protestant loyalists; protestantism; protestants
Ulster protestant revival, see evangelical revival
Ulster Special Constabulary, see 'B' Specials
Ulster Unionist Council, see Unionist
Ulster Volunteer Force 86
Unionist:
  Alliance 85
  Convention and Council 65
  Party 94–5
United Irishmen and rebellion (1798) 46, 50–2
UVF, see Ulster Volunteer Force

Vatican II 216, 217–19
Vatican Statistical Yearbook 200–1
violence 2–3, 42–5, 78, 79–80, 126–9
  see also covenant; free presbyterians; protestants; religion and 'the Troubles'; Roman catholics
Volunteers 40–1, 44–5, 47, 48

Wallis, Bruce, and Taylor 97, 106
Weber, Max 106
Wentworth, Lord (Earl Strafford) 28, 29, 30

Wesley, John 59
Westminster Confession and assembly of divines 30–1, 55–6, 119
Whigs, *see* liberal party and the liberals
William of Orange 123
Whyte, J. 6, 143, 153

Woods, Dr Michael 164–5
Workers' Party 92
Wright, Frank 120–1

Young Irelanders 80
Young Pretender 40